Lecture Notes of the Institute
for Computer Sciences, Social Informatics
and Telecommunications Engineering   194

More information about this series at http://www.springer.com/series/8197

Jason J. Jung · Pankoo Kim (Eds.)

Big Data Technologies and Applications

7th International Conference, BDTA 2016
Seoul, South Korea, November 17–18, 2016
Proceedings

 Springer

Editors
Jason J. Jung
Department of Computer Engineering
Chung-Ang University
Seoul
Korea (Republic of)

Pankoo Kim
Chosun Unversity
Gwangju
Korea (Republic of)

ISSN 1867-8211   ISSN 1867-822X (electronic)
Lecture Notes of the Institute for Computer Sciences, Social Informatics
and Telecommunications Engineering
ISBN 978-3-319-58966-4   ISBN 978-3-319-58967-1 (eBook)
DOI 10.1007/978-3-319-58967-1

Library of Congress Control Number: 2017940630

Printed on acid-free paper

This Springer imprint is published by Springer Nature
The registered company is Springer International Publishing AG
The registered company address is: Gewerbestrasse 11, 6330 Cham, Switzerland

Preface

The emergent area of big data brings together two aspects of data science and engineering: tools and applications. Theoretical foundations and practical applications of big data have set the premises for the new generation of data analytics and engineering.

BDTA 2016 was the seventh event in the series and was hosted by Chung-Ang University in Seoul, Korea, during November 17–18, 2016. BDTA 2016 was collocated with:

- The First International Workshop on Internet of Things, Social Network, and Security in Big Data (ISSB 2016)
- The First International Workshop on Digital Humanity with Big Data (DiHuBiDa 2016)

The material published in this book is divided into two main parts:

- 11 contributions by BDTA participants
- Five contributions by ISSB participants

We would like to thank to Prof. Imrich Chlamtac, the editors of the *Lecture Notes of the Institute for Computer Sciences, Social Informatics and Telecommunications Engineering* (LNICST), and the members of the Steering Committee for their kind support and encouragement in starting and continuing the BDTA Series. We deeply appreciate the efforts of the three invited speakers—Dr. Francesco Piccialli, Dr. Myunggwon Hwang, and Mr. Hyun-Jun Kim—and thank them for the interesting lectures. Special thanks also go to the organizers of ISSB2016 (Dr. Chang Choi), and DiHuBiDa2016 (O-Joun Lee). Finally, we appreciate the efforts of the local organizers on behalf of Chung-Ang University for organizing and hosting BDTA2016 and the co-located workshops.

December 2016

Jason J. Jung
Pankoo Kim

Organization

Honorary Chair

Sungjo Kim   Chung-Ang University, Seoul, South Korea

General Chairs

Jason J. Jung   Chung-Ang University, Seoul, South Korea
Pankoo Kim   Chosun University, Gwangju, South Korea

Technical Program Committee Chairs

David Camacho   Universidad Autónoma de Madrid, Spain
Paulo Novais   Universidade do Minho, Portugal
Dosam Hwang   Yeungnam University, Gyeongsan, South Korea

Website Chair

Duc Nguyen Trung   Vietnam Maritime University, Vietnam

Publicity Chair

Grzegorz J. Nalepa   AGH University of Science and Technology, Poland

Workshops Chairs

Dariusz Krol   Wroclaw University of Technology, Poland
Chang Choi   Chosun University, Gwangju, Korea
O-Joun Lee   Chung-Ang University, Seoul, South Korea

Publications Chair

Jason J. Jung   Chung-Ang University, Seoul, South Korea

Local Chair

Seungbo Park   Gachon University, Korea

Contents

BDTA

Visualization of Mixed Attributed High-Dimensional Dataset Using
Singular Value Decomposition 3
   Bindiya M. Varghese, A. Unnikrishnan, and K. Poulose Jacob

Max-flow Min-cut Algorithm in Spark with Application
to Road Networks 12
   Varun Ramesh, Shivanee Nagarajan, and Saswati Mukherjee

Association Between Regional Difference in Heart Rate Variability
and Inter-prefecture Ranking of Healthy Life Expectancy: ALLSTAR
Big Data Project in Japan. 23
   *Emi Yuda, Yuki Furukawa, Yutaka Yoshida, Junichiro Hayano, and
   ALLSTAR Research Group*

Allocation Four Neighbor Exclusive Channels to Polyhedron Clusters
in Sensor Networks 29
   ChongGun Kim, Mary Wu, and Jaemin Hong

Feature Selection Techniques for Improving Rare Class Classification
in Semiconductor Manufacturing Process 40
   Jae Kwon Kim, Kyu Cheol Cho, Jong Sik Lee, and Young Shin Han

A Novel Method for Extracting Dynamic Character Network from Movie . . . 48
   Quang Dieu Tran, Dosam Hwang, O.-Joun Lee, and Jason J. Jung

Handling Uncertainty in Clustering Art-Exhibition Visiting Styles 54
   *Francesco Gullo, Giovanni Ponti, Andrea Tagarelli, Salvatore Cuomo,
   Pasquale De Michele, and Francesco Piccialli*

Using Geotagged Resources on Social Media for Cultural Tourism:
A Case Study on Cultural Heritage Tourism 64
   Tuong Tri Nguyen, Dosam Hwang, and Jason J. Jung

Archaeological Site Image Content Retrieval and Automated Generating
Image Descriptions with Neural Network 73
   Sathit Prasomphan

Correcting Misspelled Words in Twitter Text 83
   Jeongin Kim, Eunji Lee, Taekeun Hong, and Pankoo Kim

Theoretical Concept of Inverse Kinematic Models to Determine Valid
Work Areas Using Target Coordinates from NC-Programs: A Model
Comparison to Extend a System-in-Progress as Systems Engineering Task. . .   91
    Jens Weber

ISSB

Ransomware-Prevention Technique Using Key Backup   105
    Kyungroul Lee, Insu Oh, and Kangbin Yim

Multi-level Steganography Based on Fuzzy Vault Systems
and Secret Sharing Techniques   115
    Katarzyna Koptyra and Marek R. Ogiela

On Exploiting Static and Dynamic Features in Malware Classification   122
    Jiwon Hong, Sanghyun Park, and Sang-Wook Kim

Distributed Compressive Sensing for Correlated Information Sources.   130
    Jeonghun Park, Seunggye Hwang, Janghoon Yang, Kitae Bae, Hoon Ko,
    and Dong Ku Kim

MMSE Based Interference Cancellation and Beamforming Scheme
for Signal Transmission in IoT   138
    Xin Su, YuPeng Wang, Chang Choi, and Dongmin Choi

A Multi-user-collaboration Platform Concept for Managing
Simulation-Based Optimization of Virtual Tooling as Big Data Exchange
Service: An Implementation as Proof of Concept Based on Different
Human-Machine-Interfaces   144
    Jens Weber

Author Index   155

BDTA

Visualization of Mixed Attributed High-Dimensional Dataset Using Singular Value Decomposition

Bindiya M. Varghese[✉], A. Unnikrishnan, and K. Poulose Jacob

Rajagiri College of Social Sciences, Kalamassery, India
bindiya@rajagiri.edu

Abstract. The ability to present data or information in a pictorial format makes data visualization, one of the major requirement in all data mining efforts. A thorough study of techniques, which presents visualization, it was observed that many of the described techniques are dependent on data and the visualization needs support specific to domain. On contrary, the methods based on Eigen decomposition, for elements in a higher dimensional space give meaningful depiction. The illustration of the mixed attribute data and categorical data finally signifies the data set a point in higher dimensional space, the methods of singular value decomposition were applied for demonstration in reduced dimensions (2 and 3). The data set is then projected to lower dimensions, using the prominent singular values. The proposed methods are tested with datasets from UCI Repository and compared.

Keywords: Data visualization · Mixed attribute datasets · Dimensionality reduction · SVD

1 Introduction

Visualization implies presenting data in a pictorial form. Kim Bartke states that data visualization as the plotting of data into a Cartesian space. It helps the user to have a better insight into the data. Data visualization is graphical presentation of a dataset, which provides data analysts a quality understanding of the information contents in way that is more comprehensible. The spatial representation for high dimensional data will be very handy for envisaging the relationship between the attributes.

2 Various Approaches

Sándor Kromesch et al. as geometric methods, icon-based methods, pixel-oriented techniques, hierarchical techniques, graph-based methods, and hybrid class classify the most popular visualization techniques. Geometric projection visualization techniques map the attributes to a Cartesian plane like scatter plot, or to an arbitrary space such as parallel coordinates. A matrix of scatter plots is an array of scatter plots displaying all possible pair wise groupings of dimensions or coordinates. For n-dimensional data this

© ICST Institute for Computer Sciences, Social Informatics and Telecommunications Engineering 2017
J.J. Jung and P. Kim (Eds.): BDTA 2016, LNICST 194, pp. 3–11, 2017.
DOI: 10.1007/978-3-319-58967-1_1

produces n(n − 1)/2 scatter plots with shared scales, although most often n^2 scatter plots are shown. A survey plot is a technique, which helps to extract the correlations between any two attributes of the dataset, mainly when the data is organized according to a particular dimension. Each horizontal splice in a plot relates to a particular data instance [1]. Parallel coordinate technique is a tool for envisaging multivariate data. This visualization technique maps the multi-dimensional element on to a number of axes, which are parallel. In pixel-oriented techniques the pixel representing an attribute value is colored based on the value in its domain. Recursive pattern methods orders the pixels in small clusters and arranges the clusters to form some global design [2]. The above-mentioned techniques are used to illustrate the Iris Data (UCI Repository [6]) in Figs. 1, 2, 3 and 4.

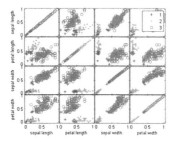

Fig. 1. Scatter matrix plot of Iris Data

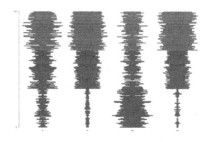

Fig. 2. Survey plot of Iris Data

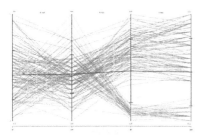

Fig. 3. Parallel coordinates of Iris Data.

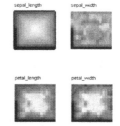

Fig. 4. Pixel oriented display of Iris Data.

3   Framework for Visualizing Mixed Attributed High-Dimensional Data

This research paper proposes a general framework to visualize a high-dimensional mixed dataset in two phases. To make it equipped for any general clustering algorithm, the mixed dataset is transformed into a uniform format. The phase includes an extension to completely categorical dataset where a frequency of the occurrence of data is made use of. A gridded representation of lower dimensional data, which was, normalized in a uniform format, is generated in the second phase. An intermediate phase requires the dimensions to be reduced.

3.1   Dataset Preprocessing

A machine-learning scheme receives an input or a set of instances, which are to be mined based on any classification, association rules or clustering techniques. The input dataset are described by the values of a set of fixed attributes after a detailed elimination of unwanted variables. To prepare this dataset adequate for data mining approaches, these data are examined for its basic data types and features. The source data is normally categorized as structured data, semi-structured or unstructured data. The structured data can be broadly classified into following types; Numeric, categorical, ordinal, nominal, ratio and interval. If dataset comprises of mixed attributes, i.e. a combination of numerical and categorical variables, then the usual approach for pre-processing is to distinctly cater different data types. Many of the clustering algorithms acts well with numeric data. The framework researched in this study starts with the various techniques to convert the categorical attributes to a numerical equivalent.

The type or magnitude of variables are not distinguished by many of the mining algorithms. The results will be affected the dominance on one particular variable over others, the algorithm treats all variables equally. Normalizing the data will eliminate this preference, thereby bringing the parameters of different units and scales into a similar plane. One of the most commonly used normalization techniques is Min-Max normalization, which performs a linear transformation on the raw data. Min-Max normalization maps a value v to v' in the range [new_min$_A$, new_max$_A$] by computing

$$v' = \frac{v - min_A}{max_A - min_A}(new\_max_A - new\_min_A) + new\_min_A \qquad (1)$$

where min_A and max_A are the minimum and maximum value of the attribute respectively. In z-score normalization, the attribute values are normalized based on the mean and the standard deviation of values. A value v is normalized to v' by calculating

$$v' = \frac{v - \mu}{\sigma}, \qquad (2)$$

where μ and σ are the mean and the standard deviation of the values respectively.

3.2   Conversion of a Mixed Dataset into a Uniform Format

Pure numeric data or on pure categorical data is apt for many of the mining algorithms. The mixed dataset in its raw form is not suitable for applying the algorithms as such. However, the real world data contains multiple types of attributes. In order to apply a data-mining algorithm, either it is required to convert the complete database to categorical or numerical in type. The proposed work is based on computing the dissimilarity computed with the co-occurrence matrices [3].

The dataset containing categorical variables is inputted as a n * p matrix, where n is the number of instances and p the number of categorical attributes. All combinations generated by p variables can be a part of the sample space. Traditional practice is to convert categorical values into a named set, for example, like low, medium and high to numeric scale. Nevertheless, certain categorical variables cannot be ordered logically like geographical data or a weapon data in crime dataset. Exploration of the relationship among the attributes may eventually help in converting the mixed set into a numerical dataset. The idea of co-occurrence is considered as a foundation in this study to find the similarities between the categorical variables.

Co-occurrence indicates that when two objects shows up frequently together, and then there is always a possibility of strong similarity between them. Closer numeric values can be assigned to those values, which is co-occurred in the dataset.

The process of pre-processing starts with the normalization of the data using any of the normalization techniques explained in the previous section. Normalization is required to eliminate the dominance of one attribute over other because of its domain size. A base attribute is selected from all categorical attribute. The selection is based on the criteria that the base item must have maximum variation of elements in the domain. The elements present in the observations of base attribute can be termed as base items. Construction of a co-occurrence matrix M of size $n * n$, where n is the number of total categorical items; m_{ij} represents the co-occurrence between item i and item j; m_{ii} represents the occurrence of item i. The similarity matrix D is given by

$$D_{xy} = \frac{|m(X, Y)|}{|m(X)| + |m(Y)| - |m(X, Y)|} \tag{3}$$

where

X the occurrence of item x;
Y the occurrence of item y;
$m(X)$ is the set of objects having the item x;
$m(X, Y)$ is the set of objects comprising both x and y.

Finally, the matrix D describes the similarity between the categorical items; higher the value, higher the similarity.

The second stage of the algorithm continues with finding a numeric feature in the same instance, which minimizes the within group variance to base attribute. The group variance can be found out by applying the following formula

$$SS_W = \sum_j \sum_i \left(X_{ij} - \bar{X}_j \right)^2 \tag{4}$$

where $\bar{X}_j$ is the mean of mapping numeric attribute of jth base item and X_{ij} is the ith value in mapping numeric attribute of jth base item. All non-base items of the categorical type can be computed by applying the following formula

$$F(x) = \sum_{j=1}^{d} a_i * v_i \tag{5}$$

where d is the number of base item; a_i is the similarity between item x and i^{th} base item taken from D_{xy}; v_i is the measured value of ith base item. Thus all the attributes in the dataset is given a numeric value.

3.3 Extension of Algorithm to Entirely Categorical Dataset

There are real world datasets, which has no or least number of numerical attributes. The given technique of finding co-occurrence and computing the group variance becomes impossible with such kind of data. To deal with the whole categorical datasets, an extension is proposed here. The first step is to add a temporary attribute FREQ to the original dataset. Each value in this column will indicate the frequency of the base item computed as per earlier methods. Specifically, each base item have its frequency in its corresponding FREQ column. The frequency is a direct indication of the strength of occurrence and is perfect to proceed with further steps.

Here a sample dataset is presented with four attributes. Attribute 2 has maximum variation of data items and hence chosen as the base item explained in the section above. The new attribute column FREQ is provisionally added to the original dataset with corresponding frequency of each base item in Attribute 2. The mean of the base items can be computed using the new FREQ column. The group variance within the base attribute can be computed as given in Eq. (4). Finally all non-base items can be given a numerical equivalent as given in Eq. (5) (Tables 1 and 2).

Table 1. Categorical dataset with four attributes

Attribute 1	Attribute 2	Attribute 3	Attribute 4
A1	B1	C1	D1
A1	B1	C2	D2
A1	B2	C1	D2
A1	B2	C2	D2
A2	B3	C3	D1
A2	B4	C3	D2
A2	B5	C3	D1

Table 2. Modified dataset with FREQ column

Attribute 1	Attribute 2	Attribute 3	Attribute 4	FREQ
A1	B1	C1	D1	2
A1	B1	C2	D2	2
A1	B2	C1	D2	2
A1	B2	C2	D2	2
A2	B3	C3	D1	1
A2	B4	C3	D2	1
A2	B5	C3	D1	1

3.4 Gridded Representation of High Dimensional Dataset

Dimensionality reduction is one of the basic functions in the steps of knowledge discovery in database is required to reduce the computational load as well as for exploratory data analysis. Each point in the observation can be characterized by n points in a n-dimensional space. There is a high possibility that this n-dimensional representation consists of sparse data. In practice, it is easier to deal with a lower dimensional dataset rather with a high dimensional dataset. However, it is mandatory to do nonlossy transformation from a higher space to a lower plane.

3.5 Dimensionality Reduction Methods

Given a set of data points $\{x_1, x_2, \ldots x_n\}$, the low-dimensional representation is

$$xi \in Rd \rightarrow yi \in Dp\,(p << d) \tag{6}$$

which preserves the information in the original dataset can be considered as a good dimensionality reduction technique Basically, there are linear or non-linear techniques for dimensionality reduction [5].

Most common Linear Dimensionality Reduction techniques are Principal Component Analysis (PCA) and Singular valued decomposition (SVD). Principal Component Analysis (PCA) replaces the original features of a data set with a smaller number of uncorrelated attributes called the principle components. If the original data set of dimension D contains highly correlated variables, then there is an effective dimensionality, d < D, explains most of the data.

Principal Component Analysis is based on Eigen value decomposition of the covariance matrix C into

$$C = PDP^T \tag{7}$$

where P is orthogonal and D is a diagonal matrix given by

$$D = Diag(\lambda_1, \lambda_2, \lambda_3 \ldots \lambda_n). \tag{8}$$

The columns of P are eigenvectors $Cx_i = \lambda x_i$ for the eigenvalues $\lambda_1 \geq \lambda_2 \geq \lambda_3 \ldots \geq \lambda_n$.

Given a complex matrix A having m rows and n columns, the matrix product U $\sum$ V is a singular value decomposition for a given matrix A if U and V, respectively, have orthonormal columns and has nonnegative elements on its principal diagonal and zeros elsewhere. i.e.

$$A_{m*n} = U_{m*r}\Sigma_{r*n}V_{n*n}^T \tag{9}$$

Where U is an orthogonal matrix, a diagonal matrix, and the transpose of an orthogonal matrix where U and V are orthogonal coordinate transformations and Σ is a rectangular-diagonal matrix of singular values. The diagonal values of Σ viz. (σ_1, $\sigma_2, \ldots, \sigma_n$) are called the singular values. The ith singular value shows the amount of variation along the ith dimension.

SVD can be used as a numerically reliable assessment of the effective rank of the matrix. The computational complexity of finding U, Σ and V in SVD when applied to a dataset of size $m * n$ (usually $m >> n$) is $4m^2n + 8mn^2 + 9n^3$.

3.6 Visualization of Dataset Using Singular Value Decomposition

The reduction of dimensions without losing the information is a greater challenge in data mining steps. While mapping a non-spatial data to 2 D plane or 3 D space, the spatial information is being added to the non-spatial component. After the initial phase of normalizing data to a range of [0 1] the dataset is applied with Singular value decomposition for dimensionality reduction. SVD when compared to PCA, it acts on the direct data matrix and the prominent singular values are taken as the principal components. It can be observed that the vectors other than the k singular values are negligible and approximate to 0. The dot product of the original data matrix to the reduced Matrix of size (n * k) is computed. When k equals 2, the matrix can be plotted to an x-y plane and when k equals 3, matrix can be plotted to space.

The concise algorithm is given below.

Input: X: of M x N size
k: dimension ; k<<N
Output: X_{reduce} with k dimensions
Steps:
1. Apply min-max normalization to normalize $X_{(M \times N)}$ into $X\text{-}Norm_{(M \times N)}$ a range of [0 1] ;

$$v' = \frac{v - min_A}{max_A - min_A}(new\_max_A - new\_min_A) + new\_min_A$$

2. [U Σ V] = SVD $(X\text{-}Norm_{(M \times N)}^T)$

3. Set principal components = first k columns of U

$$U_{reduce} = \begin{bmatrix} e_{11} & e_{12} & ...e_{1k} \\ e_{21} & e_{22} & ...e_{2k} \\ e_{n1} & e_{n2} & ...e_{nk} \end{bmatrix}_{n \times k}$$

4. $X_{reduce_{m*k}} = X^T{}_{m*n} * U_{reduce_{n*k}}$

4 Research with Various Datasets

The proposed research is tested with five major multi variate datasets from UCI repository and are discussed below. The multivariate IRIS dataset consists of 150 observations from three species of Iris: Iris setosa, Iris virginica and Iris versicolor. The length and width of petals and sepals of all three species are recorded in centimeters. Yeast is yet another multivariate set containing 1484 observations used for cellular localization sites of proteins [7] with 8 attributes. The Thyroid dataset contains 9172 instances of thyroid disease records and is provided by UCI Repository. The most frequent value technique is used to fill in the missing values. Wine dataset extracts data of 13 elements found in three types of Wine grown in the alike region of Italy. It contains 178 instances. To experiment the whole categorical case explained in the preprocessing Sect. 3.3 of this study, breast dataset with 9 categorical variables are used.

The following figures are ordered as (a) singular plot of datasets, (b) 2D plot and (c) 3D plot. Iris, Yeast, Wine, Thyroid and Breast Cancer datasets are explored in the given order from Figs. 5, 6, 7, 8 and 9.

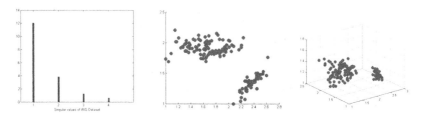

Fig. 5. a) singular plot of dataset of Iris Dataset, b) 2D plot of dataset of Iris Dataset and c) 3D plot of dataset of Iris Dataset

Fig. 6. a) singular plot of dataset of Yeast Dataset, b) 2D plot of dataset of Yeast Dataset and c) 3D plot of dataset of Yeast Dataset

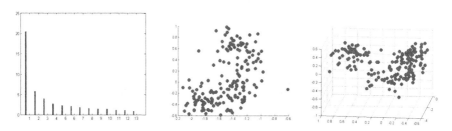

Fig. 7. a) singular plot of dataset of Wine Dataset, b) 2D plot of dataset of Wine Dataset and c) 3D plot of dataset of Wine Dataset

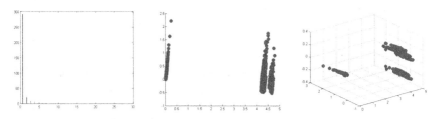

Fig. 8. a) singular plot of dataset of Thyroid Dataset, b) 2D plot of dataset of Thyroid Dataset and c) 3D plot of dataset of Thyroid Dataset

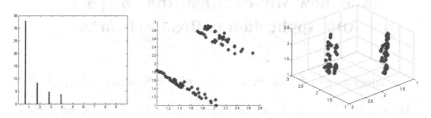

Fig. 9. a) singular plot of dataset of Breast Cancer Dataset, b) 2D plot of dataset of Breast Cancer Dataset and c) 3D plot of Breast Cancer Dataset

5 Conclusion

Existing visualization techniques are dependent on data, but an effective visualization demands the data to be independent on its type and scale. After a thorough exploration of the framework explored in this study using various multivariate datasets, Eigen decomposition or a singular value decomposition helps in representing data visually. SVD not only eliminates the curse of dimensionality, but also the pictorial mapping of a higher dimensional plane to lower dimensions, helps the user to get a better grasp on the data.

Acknowledgements. This study was conducted as a part of doctoral studies of the main author under the guidance of the co-authors and is approved by the concerned university.

References

1. Rao, R., Card, S.K.: The table lens: merging graphical and symbolic representations in an interactive focus+context visualization for tabular information. In: Proceedings of the ACM SIGCHI Conference on Human Factors in Computing Systems (1994)
2. Keim, D.A., Kriegel, H.-P., Ankerst, M.: Recursive pattern: a technique for visualizing very large amounts of data. In: Proceedings of the Visualization 1995, Atlanda (1995)
3. Shih, M.-Y., Jheng, J.-W., Lai, L.-F.: A two step method for clustering mixed categorical and numeric data. Tamkang J. Sci. Eng. **13**(1), 11–19 (2010)
4. Spence, R.: Information Visualization. Addison Wesley/ACM Press, New York (2000)
5. van der Maaten, L.J.P., Postma, E.O., van den Herik, H.J.: Dimensionality reduction: a comparative review (2008)
6. Fisher, R.A.: UCI Machine Learning Repository: Iris Data Set, January 2011
7. https://archive.ics.uci.edu/ml/index.html

Max-flow Min-cut Algorithm in Spark
with Application to Road Networks

Varun Ramesh[1(✉)], Shivanee Nagarajan[1], and Saswati Mukherjee[2]

[1] Department of Computer Science and Engineering, Anna University, Chennai, India
varun.ceg.95@gmail.com, shivanee.ceg@gmail.com
[2] Department of Information Science and Technology, Anna University, Chennai, India
msaswati@auist.net

Abstract. The max-flow min-cut problem is one of the most explored and studied problems in the area of combinatorial algorithms and optimization. In this paper, we solve the max-flow min-cut problem on large random graphs with log-normal distribution of outdegrees using the distributed Edmonds-Karp algorithm. The algorithm is implemented on a cluster using Spark. We compare the runtime between a single machine implementation and cluster implementation and analyze the impact of communication cost on runtime. In our experiments, we observe that the practical value recorded across various graphs is much lesser than the theoretical estimations primarily due to smaller diameter of the graph. Additionally, we extend this model theoretically on a large urban road network to evaluate the minimum number of sensors required for surveillance of the entire network. To validate the feasibility of this theoretical extension, we tested the model with a large log-normal graph with ~1.1 million edges and obtained a max-flow value of 54, which implies that the minimum-cut set of the graph consists of 54 edges. This is a reasonable set of edges to place the sensors compared to the total number of edges. We believe that our approach can enhance the safety of road networks throughout the world.

1 Introduction

Max-flow min-cut problem has a wide variety of applications in various domains including spam site discovery [28], community identification [11], and network optimization [1]. For instance, this problem has been previously applied on road networks to assess the reliability of the network [27]. Road networks such as that of California are extremely large and it has 1,965,206 nodes and 2,766,607 edges [24]. Protection and monitoring of such road networks for safety is a challenging task. Furthermore, to analyze such large data, it is impractical to use an extremely expensive machine equipped with voluminous storage and processing capabilities. We posit that distributed algorithm on a cluster of commodity machines, as attempted earlier [20], is an ideal solution for such large computations. Halim *et al.* in [20] used Hadoop to implement the max-flow algorithm. In similar vein, in this paper, we have chosen Spark over Hadoop to implement a distributed max-flow min-cut algorithm on a cluster.

V. Ramesh and S. Nagarajan contributed equally.

© ICST Institute for Computer Sciences, Social Informatics and Telecommunications Engineering 2017
J.J. Jung and P. Kim (Eds.): BDTA 2016, LNICST 194, pp. 12–22, 2017.
DOI: 10.1007/978-3-319-58967-1_2

Classical max-flow algorithms [14] are not feasible for real world graphs as they require the entire graph to be in memory for computation. However, max-flow algorithms such as Edmonds-Karp algorithm [10] and push-relabel algorithm [16] have good distributed settings [7,13]. A detailed explanation as to why we have chosen the Edmonds-Karp algorithm is presented in Sect. 4. We adopt the distributed algorithm from [7] and implement it on a single machine initially. Next, we implement the same on a cluster of three machines and analyze the runtime. The analysis of communication costs when iterative graph algorithms are implemented on a cluster is essential. This is because the graph will be partitioned across cluster nodes and there will be considerable communication costs amongst them to exchange information about neighboring vertices. Thus, in addition to runtime of such an algorithm, communication costs play a vital role in the overall efficiency. To the best of our knowledge, this is the first implementation of its kind on Spark where the runtime and communication costs are experimentally compared between a cluster and a single machine, based upon the max-flow min-cut algorithm.

For n vertices and m edges, the complexity as achieved in [7] is $O(cm/k) + O(cnlogk)$ with k machines and c being the max-flow value. The communication cost expected theoretically is $O(cm) + O(cnk)$. But we have noted experimentally on our cluster that the communication costs are much lesser due to smaller diameter of the graph [23] and the runtime in practice is much more efficient.

We propose a model based on our max-flow min-cut implementation to evaluate the minimum number of sensors required for surveillance of an urban road network, which is modelled as a flow graph. In a flow network, the max-flow value obtained is the minimum cut-set—the smallest total weight of edges which if removed would disconnect the source from the sink. The ratio of the cut-set to the total number of edges in the graph is then investigated to assess the feasibility of our approach. This model provides the advantage of scalability to cover the range of multiple cities, where the edge set may have over 5 million edges. An appropriate cluster size can handle such large data set and we believe that our proposal can enhance the safety of road networks throughout the world.

The rest of the paper is organized as follows. Section 2 provides background and related work. Analysis of the Edmonds-Karp algorithm and issues in the distributed settings of other algorithms are explained in Sect. 3. Various phases of the distributed algorithm are explored in Sect. 4. Our experiments on a single machine and cluster result are compared in Sect. 5, along with the investigation of practical communication costs in the cluster. Theoretical extension of the max-flow min-cut implementation to evaluate the minimum number of sensors required to provide surveillance in a large city modelled as a flow graph is performed in Sect. 6. We summarize in Sect. 7.

2 Background and Related Works

Over the years, many efficient solutions to the maximum flow problem were proposed. Some of them are the shortest augmenting path algorithm of Edmonds and Karp [10], the blocking flow algorithm of Dinic [9], the push-relabel algorithm of Goldberg and Tarjan [16] and the binary blocking flow algorithm of Goldberg and Rao [15]. The first method developed is the Ford-Fulkerson method [12]. The algorithm works by finding an augmenting path as long as one exists and then sending the minimal capacity possible through this path.

MapReduce (MR) [8] is a programming model and an implementation for processing large datasets. MapReduce programs read input data from disk, map a function across the data, reduce the results of the map, and store reduction results back in the disk. Apache Spark [29] provides programmers with an API centered on a data structure called the resilient distributed dataset (RDD), a read-only multiset of data items distributed over a cluster of machines, that is maintained in a fault-tolerant manner [30]. The availability of RDDs helps in implementing iterative algorithms as needed in our case where the dataset is visited multiple times in a loop. MapReduce has a linear dataflow pattern and Spark was developed to improve this dataflow pattern. Spark is more suitable than Apache Hadoop [19] for iterative operations because of the cost paid by Hadoop for data reloading from disk at each iteration [18].

Spark provides an API for graph algorithms that can model the Pregel [25] abstraction. Pregel was specially developed by Google to support graph algorithms on large datasets. Having mentioned about the Pregel API, it is essential that we briefly give an overview of Pregel's vertex centric approach below, and describe how it handles iterative development using supersteps.

2.1 Overview of Pregel Programming Model

Pregel computations consist of a sequence of iterations, called supersteps. During a superstep the framework invokes a user-defined function for each vertex, conceptually in parallel. The function specifies behavior at a single vertex V and a single superstep S. It can read messages sent to V in superstep $S - 1$, send messages to other vertices that will be received at superstep $S + 1$, and modify the state of V and its outgoing edges [25]. Edges are not the main focus of this model and they have no primary computation. A vertex starts in the active state and it can deactivate itself by voting to halt. It will be reactivated when it receives another message and the vertex has primary responsibility of deactivating itself again. Once all vertices deactivate and there are no more messages to be received by any of the vertices, the program will halt.

Message passing is the mode of communication used in Pregel as it is more expressive and fault tolerant than remote read. In a cluster environment, reading a value from a remote machine can incur a heavy delay. Pregel's message passing model reduces latency by delivering messages in batches asynchronously. Fault tolerance is achieved through checkpointing and a simple heartbeat mechanism is used to detect failures. MR is sometimes used to mine large graphs [21], but this can lead to suboptimal performance and usability issues [25]. Pregel has a natural graph API and is much more efficient support for iterative computations over the graph [25].

2.2 Other Related Works

General purpose distributed dataflow frameworks such as MapReduce are well developed for analyzing large unstructured data directly but direct implementations of iterative graph algorithms using complex joins is a challenging task. GraphX [17] is an efficient graph processing framework embedded within the Spark distributed dataflow system. It solves the above mentioned implementation issues by enabling composition of graphs with tabular and unstructured data. Additionally, GraphX allows users to choose either the graph or collective computational model based on the current task with no loss of efficiency [17].

Protection of Urban road networks using sensors is an important but demanding task. Applying the max-flow min-cut algorithm under the circular disk failure model to analyze the reliability of New York's road network has been performed in [27]. Placing sensors across the road network using the min-cut of the graph is explored in [3] using GNET solver. In contrast, we use our spark based max-flow min-cut implementation to place sensors efficiently across the network and discuss about practical issues such as scalability.

3 Maximum Flow Algorithm

Before we move to the distributed model, we need to understand the implications of the Edmonds-karp algorithm on a single machine setting. Additionally, we discuss issues in achieving parallelism in the push-relabel algorithm.

3.1 Overview of Max-flow Problem

A flow network [1] $G = (V, E)$ is a directed graph where each edge $(u, v) \in E$ has a non-negative capacity $c(u, v) \geq 1$. There are two end vertices in a flow network: the source vertex S and the sink vertex T. A flow is a function $f : V \times V \rightarrow R$ satisfying the following three constraints for all u, v:

1. The flow along an edge $f(u, v) \leq c(u, v)$ otherwise the edge will overflow.
2. The flow along one direction say $f(u, v) = -f(v, u)$ which is the opposite flow direction.
3. Flow must be conserved across all vertices other than the source and sink, $\sum f(u, v) = 0$ for all $(u, v) \in V - \{S, T\}$.

Augmenting path and residual graph are two very important parameters of the max-flow problem. An augmenting path is a simple path — a path that does not contain cycles. Given a flow network $G(V, E)$, and a flow $f(u, v)$ on G, we define the residual graph R with respect to $f(u, v)$ as follows.

1. The node set v of R and G are same.
2. Each edge $e = (u, v)$ of R is with a capacity of $c(u, v) - f(u, v)$.
3. Each edge $e' = (v, u)$ is with a capacity $f(u, v)$.

The push-relabel algorithm [16] maintains a preflow and converts it into maximum flow by moving flow locally between neighboring nodes using push operations. This is done under the guidance of an admissible network maintained by relabel operations. A variant of this algorithm is using the highest label node selection rule in $O(v^2 \sqrt{e})$ [4]. A parallel implementation of the push-relabel method, including some speed-up heuristics applicable in the sequential context has been studied in [13].

The push-relabel algorithm has a good distributed setting but has the main problem of low-parallelism achieved at times [22]. This means that among hundreds of machines which are computing the flow on Spark only a few of them may be working actively and further its high dependency on the heuristic function is not always suitable [5]. Moreover, other than the degenerate cases, max-flow algorithms, which make use of dynamic tree data structures, are not efficient in practice due to overhead exceeding the gains achieved [2]. On the contrary, the Edmonds-Karp algorithm works well without any heuristics and also has a high-degree of parallelism in this low diameter large graph settings. This is because of many short paths existing from the source to sink. One more primary advantage of using log-normal graphs is that they have a diameter close to $logn$ [23]. In the Edmonds-Karp algorithm mentioned below (see Algorithm 1), input is Graph $G = (V,E)$ with n vertices and m edges. $\{S,T\} \in V$ are the source and sink respectively. L is the augmenting path stored as a list. $w_1, w_2...w_n$ are the weights of the edges in the path. F_{max} is the flow variable which is incremented with the Min-flow value in every iteration.

Algorithm 1. Edmonds-karp algorithm for a single machine

1: **procedure** MAX-FLOW(S,T)
2:     **while** $augmenting path$ **do**
3:         $L \leftarrow augmenting path(S,T)$
4:         $Min flow \leftarrow min(w_1, w_2,...w_n)$ in L
5:         $Update$ $Residual$ $Graph$ and $overall$ F_{max}
6:     **return** F_{max}

We observe that for a graph G with n vertices and m edges the algorithm has a run-time complexity of $O(nm^2)$. On a single machine, if we use the bidirectional search to find the augmenting path then the time taken will be reasonable. But, if the maximum flow value is large then this will affect the runtime negatively and thus using the Pregel Shortest path approach is useful for this procedure.

4   Distributed Max-flow Model

In the distributed model, the first step is to calculate the shortest augmenting path from source to sink. The shortest augmenting path can either follow Δ-stepping method [26] or the Pregel's procedure for calculating shortest path. The Δ-stepping method has been implemented before in [6]. Given that the Pregel shortest path has an acceptable run-time, we implement the latter in Spark. After this step, we find the minimum feasible flow and broadcast this across the path which is stored as a list. The residual graph is then updated and the overall flow is incremented according to the flow value obtained.

Shortest Path using Pregel. Google's Pregel paper [25] provides with a simple algorithm to implement the single source shortest path method. In case of the max-flow problem, the S-T shortest path has to be found which is a slight variation of the single source shortest path problem.

Each vertex has three attributes - d the distance from the source S, c being the minimum capacity seen till now by that vertex and id is the identifier of the node from which the previous message had arrived. The shortest path P is obtained as output from Algorithm 2 and is stored as a list.

Algorithm 2. Shortest path using Pregel [7]

1: Make all distance value except source as ∞ and source to 0.
2: **while** $d_t = \infty$ **do**
3:     **Message sent by node i with attributes** $(d_i + 1, c_i, i)$ to each neighbor j only if $c_{ij} > 0$.
4:     **Function applied to node j upon receiving message** $(d_i + 1, c_i, i)$:
5:         $-$ set $d_j := min(d_j, d_i + 1)$
6:         $-$ if $(d_i + 1 < d_j)$
7:             $*$ set $id_j := i$
8:             $*$ $c_j := min(c_i, c_{ij})$ where c_{ij} is the capacity along edge (i, j)
9:     **Merging functions upon receiving** $m_i = (d_i + 1, c_i, i)$ and $m_j = (d_j + 1, c_j, j)$:
10:         $-$ if $d_i < d_j : m_i$
11:         $-$ else if $(d_i = d_j$ and $c_j < c_i) : m_i$
12:         $-$ else m_j

Flow Updation. The minimum capacity of the all the edges in the shortest path is the maximum flow that can be pushed through the path. The minimum value c_{min} is determined and the flow value from source to sink is incremented with c_{min}. The shortest path is also broadcasted.

Residual Graph Updation. This step includes MapReduce operation on the graph obtained from the last iteration. The key-value pair for this step is of the form $((i, j) : capacity)$.

Algorithm 3. Building Residual Graph R_G [7]

1: **Map** (input: edge,output: edge)
2: if P contains edge $(i; j) in R_G$:
3:     $-$ emit $((i, j) : c - f_{max})$
4:     $-$ emit $((j, i) : f_{max})$
5: else: emit $((i, j) : c)$ (no changes)
6: **Reduce:** sum

Runtime Analysis. For a graph with n vertices and m edges, cluster size k, the runtime complexity is mentioned as follows. For the $s - t$ shortest path algorithm, the current iteration's distance value for a node will be less than what it receives in the next iteration. Thus, only once a node emits messages outwards. For each edge we associate it with a message and thus the worst-case complexity is $O(m)$. If k machines are used in the cluster, then the complexity becomes $O(m/k)$. The step to initialize each vertex to infinity takes $O(n)$. The broadcast of the minimum flow value takes $O(nlogk)$. The next step of flow updation and residual updation in the worst case would take $O(m)$, because each edge will be passed in the iteration. Similarly, if k machines are used the worst case evaluates to $O(m/k)$. The number of iterations is bound by the maximum-flow value c. Therefore the overall run time as: $O(cm/k) + O(cn\log k)$.

Communication Cost. The communication cost of the shortest path $s - t$ method is $O(m)$. Broadcast is done to a list of at most n nodes over k machines. This leads to $O(nk)$. Similarly, the number of iterations are bound by the max-flow value c. The overall communication cost is $O(cm) + O(cnk)$. Both runtime as well as communication cost include $O(m)$ as a major factor and this will dominate the runtime of the algorithm in most cases. But this runtime is pessimistic and would rarely occur.

5   Experimental Analysis

The experimental setup is as follows. We used Spark 1.6.1 in a cluster of three machines. Each of the machine had 8 GB RAM, quad-core, intel i5 processor and ran Ubuntu 14.04 LTS. The large random graphs generated have a log-normal distribution of the outdegrees as done in [25]. The mean outdegree is 127.1, while some outliers have a degree larger than hundred thousand. Such a distribution enables the randomly generated graph to resemble real-world large-scale graphs [25].

Spark UI is used to view the timeline of the spark events, DAG visualization and real time statistics. The timeline view provides details across all jobs, within a single job and within a single stage. As these are random graphs, the results had a major dependence on the structure of the graph. Averages of results were taken for each of the test case. Runtime of the algorithm will have a high dependence on the diameter of the graph as the shortest path computation which returns a path from source to sink will scale proportionally to the diameter. A lower diameter for real-world graphs [23] will imply lower practical runtime for the max-flow min-cut algorithm implemented. In Fig. 1, single machine line plot depicts that there is an almost linear increase in runtime as the number of edges increase. Minor irregularities are observed in the single machine plot for instance between 0.18 million edges and 0.22 million edges. This is due to the decrease in the diameter of the graph which results in reduction of the runtime.

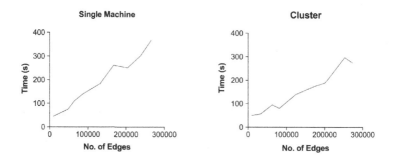

Fig. 1. Runtime analysis between single machine and cluster deployment

The cluster plot provides great insight into communication cost. It can be observed by comparing the two plots that for a similar number of edges, the cluster in most cases will provide a better runtime than a single machine as expected. But in few cases, in the initial stages, when the number of edges are about 0.05 million edges in the graph, the single machine implementation has a better runtime than the cluster. This is because communication costs prevail only in the cluster setup increasing the runtime and not in case of the single machine setup.

Further, to prove that the communication cost observed practically is much lesser than the theoretically estimated value we can look at a specific instance. For a graph with around 1.3 million edges we observed that the average runtime is about 520 s and the max-flow value is 66. Theoretically, the runtime is dominated by $O(cm)$ and for this instance it grows to $\sim 66 \times 1.3 \times 10^6$ s. This worst-case evaluation is much higher than the experimentally observed value of 520 s giving a clear indication that the worst-case bounds are very pessimistic. This creates the need for a tight upper bound which incorporates graph's diameter and topological structure.

6   Application of Distributed Max-flow Min-cut Implementation

Leading countries use public video surveillance as a major source to monitor population movements and to prevent terrorist activities. The use of Closed-circuit television (CCTV) cameras and sensors in the detection of illegal activities, data collection, recording of pre and post incident image sequences along road networks will contribute to the protection of urban population centers. Hence, it is essential for sensors to be placed in this network to detect unlawful activities. Reliability of road networks in New York city using the max-flow min-cut problem has been attempted recently [27]. Our implementation can be extended to such real world scenarios as mentioned above.

Two critical parameters in designing a system for detecting such unforeseen activities are determining how many sensors would be needed and where they should be located. We address these issues by modelling the road network as a flow graph and then the minimum cut set of the flow graph gives the optimal number of cameras or sensors to be placed.

The road network of a metropolitan city such as New York is enormous. Therefore, we process such graphs using our distributed max-flow min-cut implementation and sensors or cameras placed in the cut-set will monitor the entire network. To validate the feasibility of our approach, we evaluate the size of the min-cut and the *min-cut/edges* ratio. Smaller min-cut set implies that our approach can be practically implemented and is cost-efficient. Similar theory can be extended to a communication network to protect from a spreading virus by disconnecting the graph using the min-cut obtained for the graph.

To model the road network as a flow graph, an artificial super source and super sink node is added to the graph. Road segments are the edges and junction points between them are vertices of the graph. All edge capacities in the graph are set to unity. To facilitate the analysis, we generated a number of large graphs with log-normal distribution of outdegrees which resembles large-scale real-world road networks. An example from this set is a graph which has 11,48,734 edges and 10,000 vertices. On applying the distributed max-flow min-cut algorithm, the min-cut set obtained had 54 edges which is 0.000047 times the number of edges in the graph. This graph was solved in 462 s. Based on our results, we hypothesize that our implementation is scalable and practically applicable for large urban road networks.

To summarize, our approach evaluates the exact number of sensors required to monitor the road network and is cost-efficient when compared to placing sensors randomly across the city. The ratio of the cut-set to the edges indicates that for large real-world graphs the cut-set obtained will be a reasonable number compared to the edges and hence, this is a practically feasible solution. The largest graph tested on our cluster had more than 1.5 million edges. Larger graphs spanning over multiple cities could have in excess of 5 million edges if modelled as a flow graph. Therefore, we believe by choosing an appropriate number of machines in the cluster such large graphs can be easily processed in reasonable time.

7   Conclusion

In this paper, we implemented the distributed version of the Edmonds-Karp algorithm using Spark. In prior efforts only single machine analysis have been performed experimentally but do not include communication cost observations over a cluster for the max-flow algorithm on Spark. We believe this is the first attempt where the runtime and the communication cost based upon the maximum flow algorithm, have been compared across a single machine and a cluster experimentally. Communication cost is observed to have a considerable impact on the runtime. We then propose a method of evaluation and placement of sensors in a large city with millions of road segments, which is modelled as a flow graph. The cut-set of a log-normal graph with 11,48,734 edges is 0.000047 times smaller than the total number of road segments. Thus, our experiments show a promising and scalable approach to place sensors in a large urban road networks based on the distributed max-flow min-cut implementation.

References

1. Ahuja, R.K., Magnanti, T.L., Orlin, J.B.: Network flows: theory, algorithms, and applications (1993)
2. Badics, T., Boros, E.: Implementing a maximum flow algorithm: experiments with dynamic trees. Netw. Flows Matching First DIMACS Implement. Chall. **12**, 43 (1993)
3. Barnett, R.L., Sean Bovey, D., Atwell, R.J., Anderson, L.B.: Application of the maximum flow problem to sensor placement on urban road networks for homeland security. Homel. Secur. Aff. **3**(3), 1–15 (2007)
4. Cheriyan, J., Maheshwari, S.N.: Analysis of preflow push algorithms for maximum network flow. SIAM J. Comput. **18**(6), 1057–1086 (1989)
5. Cherkassky, B.V., Goldberg, A.V.: On implementing the push-relabel method for the maximum flow problem. Algorithmica **19**(4), 390–410 (1997)
6. Crobak, J.R., Berry, J.W., Madduri, K., Bader, D.A.: Advanced shortest paths algorithms on a massively-multithreaded architecture. In: 2007 IEEE International Parallel and Distributed Processing Symposium, pp. 1–8. IEEE (2007)
7. Dancoisne, B., Dupont, E., Zhang, W.: Distributed max-flow in spark (2015)

8. Dean, J., Ghemawat, S.: MapReduce: simplified data processing on large clusters. Commun. ACM **51**(1), 107–113 (2008)
9. Dinic, E.A.: Algorithm for solution of a problem of maximum flow in a network with power estimation. Sov. Math. Dokl. **11**(5), 1277–1280 (1970)
10. Edmonds, J., Karp, R.M.: Theoretical improvements in algorithmic efficiency for network flow problems. J. ACM (JACM) **19**(2), 248–264 (1972)
11. Flake, G.W., Lawrence, S., Giles, C.L.: Efficient identification of web communities. In: Proceedings of the sixth ACM SIGKDD International Conference on Knowledge Discovery and Data Mining, pp. 150–160. ACM (2000)
12. Ford, L.R., Fulkerson, D.R.: Maximal flow through a network. Can. J. Math. **8**(3), 399–404 (1956)
13. Goldberg, A.V.: Efficient graph algorithms for sequential and parallel computers. Ph.D. thesis, Massachusetts Instutute of Technology, Department of Electrical Engineering and Computer Science (1987)
14. Goldberg, A.V.: Recent developments in maximum flow algorithms. In: Arnborg, S., Ivansson, L. (eds.) SWAT 1998. LNCS, vol. 1432, pp. 1–10. Springer, Heidelberg (1998). doi:10.1007/BFb0054350
15. Goldberg, A.V., Rao, S.: Beyond the flow decomposition barrier. J. ACM (JACM) **45**(5), 783–797 (1998)
16. Goldberg, A.V., Tarjan, R.E.: A new approach to the maximum-flow problem. J. ACM (JACM) **35**(4), 921–940 (1988)
17. Gonzalez, J.E., Xin, R.S., Dave, A., Crankshaw, D., Franklin, M.J., Stoica, I.: GraphX: graph processing in a distributed dataflow framework. In: 11th USENIX Symposium on Operating Systems Design and Implementation (OSDI 2014), pp. 599–613 (2014)
18. Lei, G., Li, H.: Memory or time: performance evaluation for iterative operation on Hadoop and Spark. In: 2013 IEEE 10th International Conference on High Performance Computing and Communications & 2013 IEEE International Conference on Embedded and Ubiquitous Computing (HPCC_EUC), pp. 721–727. IEEE (2013)
19. Apache Hadoop: Hadoop (2009)
20. Halim, F., Yap, R.H., Yongzheng, W.: A MapReduce-based maximum-flow algorithm for large small-world network graphs. In: 2011 31st International Conference on Distributed Computing Systems (ICDCS), pp. 192–202. IEEE (2011)
21. Kang, U., Tsourakakis, C.E., Faloutsos, C.: PEGASUS: a peta-scale graph mining system implementation and observations. In: 2009 Ninth IEEE International Conference on Data Mining, pp. 229–238. IEEE (2009)
22. Kulkarni, M., Burtscher, M., Inkulu, R., Pingali, K., Casçaval, C.: How much parallelism is there in irregular applications? In: ACM Sigplan Notices, vol. 44, pp. 3–14. ACM (2009)
23. Leskovec, J., Kleinberg, J., Faloutsos, C.: Graphs over time: densification laws, shrinking diameters and possible explanations. In: Proceedings of the Eleventh ACM SIGKDD International Conference on Knowledge Discovery in Data Mining, pp. 177–187. ACM (2005)
24. Leskovec, J., Krevl, A.: SNAP Datasets: Stanford large network dataset collection, June 2014. http://snap.stanford.edu/data
25. Malewicz, G., Austern, M.H., Bik, A.J., Dehnert, J.C., Horn, I., Leiser, N., Czajkowski, G.: Pregel: a system for large-scale graph processing. In: Proceedings of the 2010 ACM SIGMOD International Conference on Management of data, pp. 135–146. ACM (2010)
26. Meyer, U., Sanders, P.: δ-stepping: a parallelizable shortest path algorithm. J. Algorithm. **49**(1), 114–152 (2003)
27. Otsuki, K., Kobayashi, Y., Murota, K.: Improved max-flow min-cut algorithms in a circular disk failure model with application to a road network. Eur. J. Oper. Res. **248**(2), 396–403 (2016)

28. Saito, H., Toyoda, M., Kitsuregawa, M., Aihara, K.: A large-scale study of link spam detection by graph algorithms. In: Proceedings of the 3rd International Workshop on Adversarial Information Retrieval on the Web, pp. 45–48. ACM (2007)
29. Apache Spark: Apache spark^{TM} is a fast and general engine for large-scale data processing (2016)
30. Zaharia, M., Chowdhury, M., Franklin, M.J., Shenker, S., Stoica, I.: Spark: cluster computing with working sets. HotCloud **10**, 10 (2010)

Association Between Regional Difference in Heart Rate Variability and Inter-prefecture Ranking of Healthy Life Expectancy: ALLSTAR Big Data Project in Japan

Emi Yuda, Yuki Furukawa$^{(\boxtimes)}$, Yutaka Yoshida,
Junichiro Hayano, and ALLSTAR Research Group

Department of Medical Education, Nagoya City University Graduate School of Medical Sciences,
Mizuho-cho, Mizuho-ku, Nagoya 467-8602, Japan
{emi21,yyoshida,hayano}@med.nagoya-cu.ac.jp,
furukawa.yuki@gmail.com

Abstract. As a physiological big-data project named ALLSTAR, we have developed a 24-hr ambulatory electrocardiogram database of 81,615 males and 103,038 females ($\geq$20 yr) from all over Japan. With this database, we examined if regional differences in heart rate (HR) and HR variability (HRV) are associated with the inter-prefecture rankings of healthy life expectancy (HALE) and of average life expectancy (ALE) in Japan. According to reports by the Japanese Ministry of Health, Labour and Welfare (2013), subjects in each sex were grouped into short, middle, and long HALE and ALE tertiles by their living prefectures. Standard deviation of 24-h normal-to-normal R-R intervals (SDNN) increased progressively with increasing HALE tertiles in both sexes (Ps < 0.001), while it showed no consistent associations with ALE. Conversely, HR decreased progressively with increasing ALE tertiles in females (P < 0.001), while it showed no consistent association with HALE. These suggest HRV may reflect a biological property relating to long HALE.

Keywords: Allostatic State Mapping by Ambulatory ECG Repository (ALLSTAR) · Physiological big data · Healthy life expectancy · Heart rate variability

1 Introduction

Healthy life expectancy (HALE) is defined as the period in the life when people might live without restriction of their daily activities due to health problems. According to reports by the Japanese Ministry of Health, Labour and Welfare (2013) [1], HALE of Japanese men and women are 71.19 and 74.21 yr, respectively. Because their average life expectancy (ALE) are reported to be 80.21 and 86.61 yr, respectively, there are gaps of 9.02 yr for men and 12.40 yr for women, which detract individual quality of life, reduce the activities of society, and cause a substantial social burden. Shortening of this gap is currently one of the most urgent challenges to mankind. In order to address to this problem effectively, however, it seems important to clarify the biological properties that determine HALE separately from ALE.

© ICST Institute for Computer Sciences, Social Informatics and Telecommunications Engineering 2017
J.J. Jung and P. Kim (Eds.): BDTA 2016, LNICST 194, pp. 23–28, 2017.
DOI: 10.1007/978-3-319-58967-1_3

As a physiological big-data building project named Allostatic State Mapping by Ambulatory ECG Repository (ALLSTAR) [2], we have developed a database of 24-h ambulatory electrocardiograms (ECGs) collected from all over Japan since 2007 and currently, the size has reached about 240 thousand cases. All data are associated with date and geographic location (postal code) of recording, which allow us to analyze the regional differences in ECG-derived indices such as 24-h heart rate (HR) and heart rate variability (HRV) [3]. The above mentioned reports by the Japanese Ministry of Health, Labour and Welfare (2013) [1] have also shown the presence of regional differences in HALE and ALE and provided their inter-prefecture rankings for each sex. In the present study, we combined our ECG big data and these databases and examined if regional differences in HR and HRV are associated with the inter-prefecture ranking of HALE and ALE.

2   Methods

2.1   ECG Data

We used 24-h Holter ECG big data of ALLSTAR project [2]. The data were collected between November 2007 and July 2014 at three ECG analysis centers (Sapporo, Tokyo, and Nagoya) in Japan. We used data only from subjects aged ≥ 20 yr who have agreed with the usage of their data for this study. The study protocol has been approved by the Research Ethics Committee of Nagoya City University Graduate School of Medical Sciences (No. 709).

The inclusion criteria of data for this study were those of subjects aged ≥ 20 yr, record length ≥ 21.6 h (90% of 24 h), and ECG showing sinus rhythm (normal heart rhythm) ≥ 19.2 h (80% of 24 h). Exclusion criteria were ECG data from patients with cardiac pace maker and those showing persistent or paroxysmal atrial fibrillation.

2.2   Analysis of HRV

Holter ECG recordings were processed according to the standard method [3]. Briefly, ECG recordings were analyzed with an ambulatory ECG scanner (Cardy Analyzer 05, Suzuken Co., Ltd., Nagoya, Japan) by which QRS complexes were detected and labeled automatically. QRS complexes were considered as sinus rhythm only when (1) they had a narrow supraventricular morphology, (2) R-R intervals were between 300 and 2000 ms and differed $\leq 20\%$ from the average of 5 preceding sinus rhythm R-R intervals, and (3) consecutive R-R interval differences were ≤ 200 ms. Results of the automatic analysis were reviewed and any errors in R-wave detection and in QRS waveform labeling were edited manually.

Time series of 24-h R-R intervals data were analyzed for HRV measures according to the recommendations by the Task Force of the European Society of Cardiology and the North American Society of Pacing and Electrophysiology [3]. Mean HR was calculated from the 24-h average of normal-to-normal R-R intervals consisting of two consecutive QRS complexes in sinus rhythm and standard deviation of normal-to-normal R-R interval over 24-hr (SDNN) [4] was calculated.

2.3   Statistical Analysis

The subjects in ALLSTAR database were classified into 47 prefectures according to the postal code of their recording locations. Then, the prefectures were sorted by the inter-prefecture rankings of HALE and ALE in each sex and were further categorized into tertiles of short, middle, and long HALE and ALE, so that the number of subjects becomes as equivalent as possible among the tertiles.

We used SAS 9.4 software (SAS Institute, Cary, NC) for the statistical analysis. Differences in HR and SDNN among tertiles of HALE and ALE were evaluated by ANCOVA with General Linear Model procedure adjusted for the effects of age in each sex. Bonferroni correction was used for post-hoc multiple comparisons between tertiles. We used $\alpha < 0.001$ to guard against type 1 statistical error.

3   Results

According to the inclusion and exclusion criteria, 24-hr ambulatory ECG data in 79,354 men (median age [IQR], 65 [56–76] yr) and 99,961 women (67 [59–78] yr) were extracted for this study. They were categorized into HALE and ALE tertiles by their inter-prefecture rankings of their living prefectures in each sex. Table 1 shows the number of subjects and the range of HALE and ALE in each tertile.

Table 1. Subjects grouped into tertiles according to inter-prefecture rankings of healthy life expectancy (HALE) and average life expectancy (ALE) of their living prefectures by sex

Tertile	HALE		ALE	
	N	Range (yr)	N	Range (yr)
Male				
Short	24,872	68.9–69.9	25,687	77.3–79.6
Middle	28,969	70.0–70.8	15,229	79.7–79.8
Long	25,513	70.9–71.7	38,438	79.9–81.0
Female				
Short	34,650	72.3–73.0	32,922	85.4–86.2
Middle	31,391	73.1–73.7	27,796	86.3–86.5
Long	33,920	73.8–75.3	39,243	86.6–87.2

ANCOVA revealed that 24-h SDNN (SDNN) increased progressively with increasing HALE tertiles for both sexes (P < 0.001 for both), while no consistent associations was observed between 24-hr mean HR and HALE tertiles in either sexes (Fig. 1). Conversely, mean HR decreased progressively with increasing ALE tertiles only in women (P < 0.001), while mean HR in men and SDNN in both sexes showed no consistent association with ALE tertiles. The same results were obtained even when the analyses were performed separately in subjects with age <70 yr (45,271 males and 50,763 females) and in those ≥70 yr (34,083 males and 49,198 females).

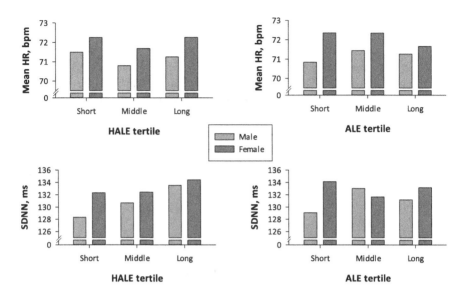

Fig. 1. Mean heart rate (HR) and standard deviation of normal-to-normal R-R intervals (SDNN) over 24 h in the tertiles of inter-prefecture rankings of healthy life expectancy (HALE) and average life expectancy (ALE) in each sex.

4   Discussion

To our knowledge, this is the first study to report the association between HRV and HALE. We found that reginal difference in HRV measured as SDNN is associated with inter-prefecture ranking of HALE but not with the ranking of ALE. We observed that SDNN increased progressively in the order of short, middle, and long HALE tertiles, while it showed no consistent association with ALE tertiles. Conversely, mean HR decreased with increasing ALE tertiles in women, while it showed no consistent association with HALE tertiles. These findings suggest that HR and HRV may involve differently in HALE and ALE and that HRV may reflect a biological property relating to long HALE.

As aging of population is progressing in the global scale, the paradigm of medicine is shifting from the extension of ALE to the shortening the gap between ALE and HALE. Unlike the causes of death that shorten ALE, the major factor that shortens HALE is thought to be the decay and loss of mental and physical activities, which creep unconsciously under daily life. To cope with this issue, not only conventional medical approaches such as preventions, rescues, and therapies of diseases but also the early detections of activity decay and the self-management of biological or behavioral properties that are associated with early/rapid development of mental and physical inactivation seem essential. In this aspect, the findings of this study seem to provide an important insight into this issue, suggesting the association of HRV with HALE.

Although the mechanisms of the association between HRV and HALE cannot be determined by the present study because of its observational nature, some speculations may be possible. First, because the significant association between SDNN and HALE was observed even when the analysis was performed only in subjects with age <70 yr, the association may not be attributable to the results of a greater proportion of people who had already lost their healthy life in the short and middle tertiles of HALE. If this is true, lower SDNN might be a predictor of early/rapid development of inactivity that could lead to short HALE. Second, in a previous study of ALLSTAR big-data project, we have reported that regional difference in physical activity is associated with inter-prefecture ranking of HALE [5]. In this previous study, we have assessed physical activity by 3-axis accelerometer installed in Holter ECG recorders in 18,875 men and 23,541 women. We observed that physical activity level is increasing progressively with increasing tertiles of HALE. This suggests that the maintenance of physical activity may be a factor promoting longer HALE. By contrast, because SDNN is known to reflect mainly parasympathetic activity [3] and 24-h SDNN is mostly determined by SDNN during nighttime [6]. Thus, the findings of the present study combined with those of our previous study lead to the speculation that the maintenance or enhancement of rest-activity rhythm in daily life may be an important factor for long HALE.

The present study has several limitations. First, the big data were collected from patients who underwent Holter ECG for certain clinical reason(s). Thus, we need to consider the presence of sampling bias, although it would cause common effects on data in all prefectures. Second, although the number of subjects was about 200 thousand, it may still not be sufficient to represent the regional characteristics. This seems critical to the indirect analyses of the relationships between sampled observations (HR and HRV) and population characteristics (HALE and ALE). Expansion of the size of database and inclusion of healthy people's data seem important to confirm the findings in the future studies.

5   Conclusions

As ALLSTAR big-data project, we have developed a database of 24-hr ambulatory ECG from all over Japan. With this database, we analyzed the associations between regional differences in HR and HRV and the inter-prefecture rankings of HALE and ALE. We found that HRV measured as SDNN increases progressively in short, middle, and long HALE prefectures, while it showed no consistent associations with ALE. Our findings suggest that HRV may reflect a biological property relating to long HALE.

References

1. Progress of each goal in Healthy Japan 21 (2nd stage), the Ministry of Health, Welfare, and Labour.   http://www.mhlw.go.jp/file/05-Shingikai-10601000-Daijinkanboukouseikagakuka-Kouseikagakuka/sinntyoku.pdf
2. Allostatic State Mapping by Ambulatory ECG Repository (ALLSTAR) project. http://www.med.nagoya-cu.ac.jp/mededu.dir/allstar/
3. Camm, A.J., Malik, M., Bigger, Jr., J.T., Breithardt, G., Cerutti, S., Cohen, R.J., Coumel, P., Fallen, E.L., Kleiger, R.E., Lombardi, F., Malliani, A., Moss, A.J., Rottman, J.N., Schmidt, G., Schwartz, P.J., Singer, D.H. Task Force of the European Society of Cardiology and the North American Society of Pacing and Electrophysiology: Heart rate variability: standards of measurement, physiological interpretation and clinical use. Circulation **93**, 1043–1065 (1996)

4. Kleiger, R.E., Miller, J.P., Bigger Jr., J.T., Moss, A.J.: Decreased heart rate variability and its association with increased mortality after acute myocardial infarction. Am. J. Cardiol. **59**(4), 256–262 (1987)
5. Yuda, E., Yoshida, Y., Hayano, J., ALLSTAR Research Group: Regional Difference in Physical Activity is Associated with the Ranking of Healthy Life Expectancy among Prefectures in Japan: The 78th National Convention of IPSJ, 10–12 March 2016
6. Malik, M., Farrell, T., Camm, A.J.: Circadian rhythm of heart rate variability after acute myocardial infarction and its influence on the prognostic value of heart rate variability. Am. J. Cardiol. **66**, 1049–1054 (1990)

Allocation Four Neighbor Exclusive Channels to Polyhedron Clusters in Sensor Networks

ChongGun Kim[2], Mary Wu[1(✉)], and Jaemin Hong[2]

[1] Department of Computer Culture, Yongnam Theological University and Seminary,
Gyeongsan, Korea
mrwu@ynu.ac.kr
[2] Department of Computer Engineering, Yeungnam University, Gyeongsan, Korea
cgkim@yu.ac.kr, hjm4606@naver.com

Abstract. In sensor networks, due to the limited resources of sensor nodes, energy efficiency is an important issue. Clustering reduces communication between sensor nodes and improves energy efficiency by transmitting the data collected from cluster members by a cluster header to a sink node. Due to the frequency characteristics in a wireless communication environment, interference and collision may occur between neighboring clusters, which may lead causing network overhead and increasing energy consumption due to the resulting re-transmission. Inter-cluster interference and collisions can be resolved by allocating non-overlapping channels between neighboring clusters. In our previous study, a channel allocation algorithm is proposed. It allocates exclusive channels among neighbor clusters in various polygonal cluster networks. But, we have found that channel collision occurs among neighbor clusters in the process of applying the algorithm in complex cluster topologies. In order to complete the non-collision channel allocation, a straightforward algorithm which has not completed in the previous studies is proposed. We provide 4 rules for proving the proposed algorithm has no weakness. The result of experiment shows that our proposed algorithm successfully allocates exclusive channels between neighboring clusters using only 4 channels in a very complex cluster topology environment.

Keywords: Energy efficiency · Clustering · Inter-cluster interference · Non-overlapping channels · Four-color theorem

1 Introduction

Wireless sensor networks collect environmental data and are applied for various purposes such as intrusion detection in environmental monitoring of temperature and humidity, military areas, and area security. Sensor nodes become aware of the surrounding symptoms and transmit the measured data to sink nodes, which in turn analyzes the data. Due to the limited energy resources of sensor nodes, limitations arise on the wireless sensor networks. Many studies on the efficient use of energy have been done to overcome this problem [1–7]. In general, neighboring sensor nodes collect similar information, leading to large energy wastage due to duplicate transmission of similar information. As a result, many cluster methods on sensor networks have been

© ICST Institute for Computer Sciences, Social Informatics and Telecommunications Engineering 2017
J.J. Jung and P. Kim (Eds.): BDTA 2016, LNICST 194, pp. 29–39, 2017.
DOI: 10.1007/978-3-319-58967-1_4

studied. Clustering, divides the sensor network into groups of nodes, is an effective method for achieving a high level of energy efficiency. In clustering, each node becomes a member of the cluster and the cluster headers aggregate data collected from members and then transmits it to a sink node. This prevents duplicated transmission of similar information and increases the energy efficiency of the sensor network [5–7].

Transmission synchronization between clusters can be done through non-overlapping channel assignments between neighboring clusters. Figure 1 shows the examples of channel reuse rate 4 and 7 [8, 9].

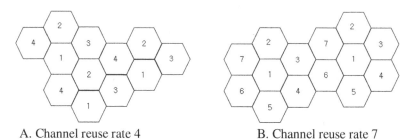

A. Channel reuse rate 4   B. Channel reuse rate 7

Fig. 1. Reuse rate in hexagonal models

In the Low-Energy Adaptive Clustering Hierarchy (LEACH) [1], representative clustering protocol for sensor networks, Transmission synchronization between nodes in clusters uses TDMA. Nodes located near the cluster boundary may interfere with the data transmission of another cluster node. Figure 2 shows that the transmission of the node which is located near the cluster boundary can interfere in data transmissions of the other clusters.

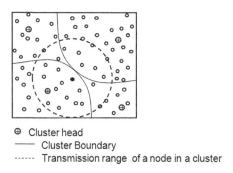

⊕ Cluster head
── Cluster Boundary
------ Transmission range of a node in a cluster

Fig. 2. Interference in cluster sensor networks

For inter-cluster synchronization in LEACH, each cluster communicates using different CDMA codes [1]. It is inefficient to implement CDMA code on low cost sensor nodes at cost.

TDMA-based Avoiding Collision (TAC) [10] uses TDMA to resolve inter-cluster synchronization. Cluster are assigned a different group ID between adjacent clusters and the group ID ranges from 1 to 4. The Initial cluster broadcasts a group allocation message including its own group ID. The clusters which receive it allocate the group ID which isn't same group ID. This process is continues until all clusters have group IDs in the wireless sensor network. This causes many control messages and delays for synchronization in the sensor network.

In [11, 12], an exclusive channel allocation based on matrices has been proposed. It assigns different channels between neighboring clusters based on the matrices, such as an adjacency, a topology and a resource allocation. The complicate calculations which are required for the exclusive channel allocation are made by a server with non-limited resource. Therefore, it doesn't require the exchange of many messages among sensor nodes. But this method is applicable in a hexagonal cluster topology, it may occur collisions among neighboring clusters in various cluster topologies.

In our previous study [13], the channel allocation method based on the four-color theorem [14–18] has been proposed. It allocates different channels among neighbor clusters in various cluster networks. It is very intuitive and simple to apply as compared to the previous methods which prove the four-color theorem. But, we found that channel collision occurs between neighbor clusters in complex cluster topologies. In order to complete the exclusive channel allocation algorithm which has not completed in the previous study, a local channel allocation method is studied [19]. But this process is very complicated.

In this paper, we propose straightforward algorithm to allocate exclusive channels in the neighbor clusters. The experiment shows that the enhanced algorithm successfully assigns non-overlapping channels between neighbors using only four channels in a very complicate cluster topology.

2   The Concept of Exclusive Channel Allocation

Non-overlapping channels among adjacent clusters can be assigned by only using 4 channels in some form of topology based on the four-color theorem. Figure 3(b) shows the result of the exclusive channel allocation using the four channel numbers 1, 2, 3, 4 in the topology of Fig. 3(a).

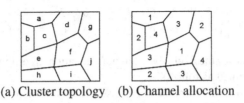

(a) Cluster topology   (b) Channel allocation

Fig. 3. Cluster topology and the result of exclusive channel allocation

For non-overlapping channel allocation among neighbors in the cluster sensor network, we use the adjacency matrix A which presents the adjacency relation among clusters and the exclusive channel matrix EC which presents the result of exclusive channel allocation based on the matrix A.

(1) shows the A matrix of the cluster topology of Fig. 3(a). In the A, when two clusters are adjacent, the elements are represented by 'n' and two clusters are not adjacent the elements are represented by '0' [20]. (2) shows the EC matrix of the channel allocation of Fig. 3(b).

$$
A = \begin{array}{c} \\ a \\ b \\ c \\ d \\ e \\ f \\ g \\ h \\ i \\ j \end{array}
\begin{array}{c}
\begin{array}{cccccccccc} a & b & c & d & e & f & g & h & i & j \end{array} \\
\left[\begin{array}{cccccccccc}
n & n & n & n & 0 & 0 & 0 & 0 & 0 & 0 \\
n & n & n & 0 & n & 0 & 0 & 0 & 0 & 0 \\
n & n & n & n & n & 0 & 0 & 0 & 0 & 0 \\
n & 0 & n & n & 0 & n & n & 0 & 0 & 0 \\
0 & n & n & 0 & n & n & 0 & n & 0 & 0 \\
0 & 0 & 0 & n & n & n & n & 0 & n & n \\
0 & 0 & 0 & n & 0 & n & n & 0 & 0 & n \\
0 & 0 & 0 & 0 & n & 0 & 0 & n & n & 0 \\
0 & 0 & 0 & 0 & 0 & n & 0 & n & n & n \\
0 & 0 & 0 & 0 & 0 & n & n & 0 & n & n
\end{array}\right].
\end{array}
\tag{1}
$$

The following EC shows the exclusive channel allocation matrix. In case of cluster a, the assigned channel is 1 and the neighbor channels are 2 in cluster b, 4 in cluster c and 3 in cluster d.

$$
EC = \begin{array}{c} \\ a \\ b \\ c \\ d \\ e \\ f \\ g \\ h \\ i \\ j \end{array}
\begin{array}{c}
\begin{array}{cccccccccc} a & b & c & d & e & f & g & h & i & j \end{array} \\
\left[\begin{array}{cccccccccc}
1 & 2 & 4 & 3 & 0 & 0 & 0 & 0 & 0 & 0 \\
1 & 2 & 4 & 0 & 3 & 0 & 0 & 0 & 0 & 0 \\
1 & 2 & 4 & 3 & 3 & 0 & 0 & 0 & 0 & 0 \\
1 & 0 & 4 & 3 & 0 & 1 & 2 & 0 & 0 & 0 \\
0 & 2 & 4 & 0 & 3 & 1 & 0 & 2 & 0 & 0 \\
0 & 0 & 0 & 3 & 3 & 1 & 2 & 0 & 3 & 4 \\
0 & 0 & 0 & 3 & 0 & 1 & 2 & 0 & 0 & 4 \\
0 & 0 & 0 & 0 & 3 & 0 & 0 & 2 & 3 & 0 \\
0 & 0 & 0 & 0 & 0 & 1 & 0 & 2 & 3 & 4 \\
0 & 0 & 0 & 0 & 0 & 1 & 2 & 0 & 3 & 4
\end{array}\right].
\end{array}
\tag{2}
$$

The following shows the exclusive channel allocation algorithm. The notations for the algorithm are defined as follows;

The channel allocation formula (3) is used to assign non-overlapping channels among adjacent clusters. The channel allocation number is used in the range of from 1 to 4.

$$(3i + 2j) \% 4 + 1 \tag{3}$$

i, j are the channel numbers of the two reference clusters.

3   Channel Allocation Algorithm

<Rule 1> A non-overlapping channel can be allocated to a polygon cluster which has less than 3 allocated neighbors with different channel in the four-channel allocation problem at the polygonal cluster networks.

<Rule 2> To allocate non-overlapping channel in the entire area of a polygonal network, the channel allocated area of is extended by step by step. The boundary of the channel allocated area (i step, i >= 2) is adjacent less than 2-hop from the previous boundary. The minimum polygon is triangle, so un-allocated clusters which have adjacent more than 3 channel allocated neighbors can be avoided by this rule.

<Rule 3> In the extending of channel allocated area, a cluster of the 2-hop neighbors from the previous boundary can be selected as the followings; a cluster which already has 3 allocated neighbor is selected. By applying this rule to all adjacent clusters from the previous boundary, a new boundary of expanded allocated area is obtained.

<Rule 4> During the process of allocating channels in candidate clusters based on previous rules, a cluster which adjust sides to the allocated area boundary with more than 4 already allocated clusters can be appeared, then by switching channels between neighbour clusters, exclusive channels can be allocated among neighbours. Especially, when a cluster has aside 4 already channel allocated clusters, then the situation can be solved by changing with a channel that was allocated to the before procedure.
   Exclusive Channel allocation algorithm is shown as follows;

1. Select one of cluster which has most many sides in the polygons as the start point and apply a channel.
2. In the clusters which are attached to already allocated area, a cluster which is attached most many sides is selected as a start point and non-conflictive channel with neighbours is allocated. In this case, most many sides of a cluster adjusted with the allocated area may be less than 3. Then select the next candidate cluster in clockwise or count-clockwise directions.
3. In each area expansion iteration, the channel allocated area must be expanded. The width of expansion is more than one cluster to outward from the allocated area.
4. All of the outward clusters those are adjusted with expanding allocated area must have not more than 2 adjusted sides, then an iteration of expanding allocated area is completed.
5. During an expanding allocated area, a case that an allocating candidate neighbour which may have more than 4 adjust sides from the allocated area is happen, then the candidate neighbour treated earlier than present cluster channel allocation or a local channel exchanging among clusters must be needed.

4 Network Operation

The cluster headers perform a central role in controlling networking such as synchronization, data collection, data transfer, so they consume a large amount of energy compared to member nodes. Before the energy of the cluster headers is exhausted, the selection process of new cluster headers is required. Therefore, clustering is performed periodically or re-clustering is required depending on the amount of energy remaining in the cluster header.

The operation is divided into rounds, as in LEACH [1]. Figure 4 shows the configuration of a round.

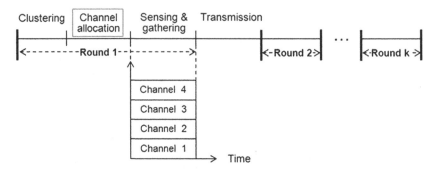

Fig. 4. Round configuration

The proposed algorithm is performed in a system with no constraints on system resources such as power, energy, memory, and processing capability.

In the channel allocation, cluster headers broadcast a hello message, The cluster headers which receive it recognize the existence of the neighbor cluster and generate the list of their neighbors and send it to a gateway or a server. The system performs a channel assignment algorithm for non-overlapping channel assignment.

5 Experiment

Performance evaluation experiments of the proposed channel assignment algorithm were performed. Figures 5, 6, 7, 8 and 9 shows that the process of channel allocation using the proposed algorithm. The numbers indicated in squares show allocation sequence. The numbers indicated in circles show the number of channels assigned to each cluster.

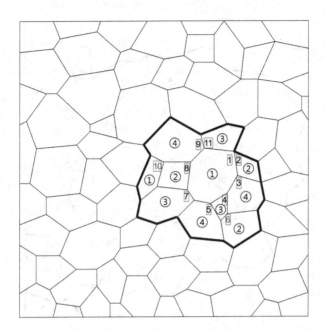

Fig. 5. Iteration result for the exclusive channel allocation area expansion (Color figure online)

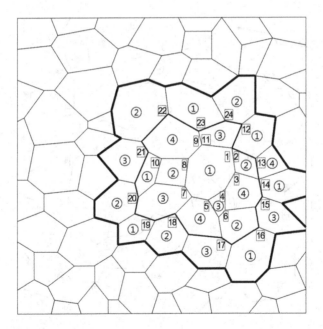

Fig. 6. Iteration result for the exclusive channel allocation area expansion

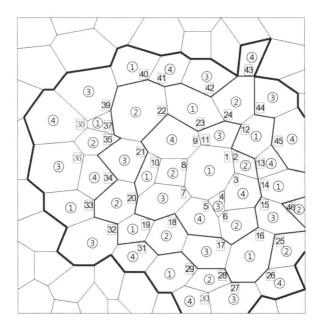

Fig. 7. Iteration result for the exclusive channel allocation area expansion (Color figure online)

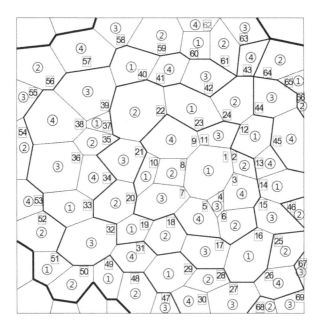

Fig. 8. Iteration result for the exclusive channel allocation area expansion (Color figure online)

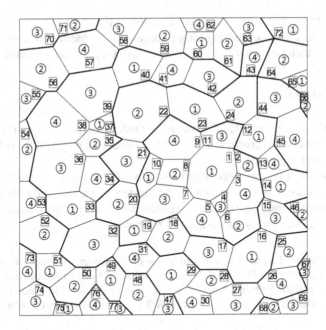

Fig. 9. Iteration result for the exclusive channel allocation area expansion

Figure 5 shows the first step of the channel allocation process. The cluster which has 8 neighbors is selected, and the channel 1 is assigned to the cluster. In the first step of the channel allocation, the channel which does not overlap is allocated to the one-hop neighbor clusters of the selected cluster. The channel 2 is assigned to one of the one-hop neighbor clusters of the first cluster. The channel 4 is allocated to the third cluster, based on the two clusters which are the first and the second cluster by the formula (3). This channel allocation process is continuously repeated based on the first cluster and the previous cluster allocated. After the fifth channel allocation, a non-overlapping channel is assigned to the cluster which has 3 neighbors to which channels are already assigned. The cluster isn't one-hop neighbor of the first cluster, but, a non-overlapping channel is allocated by rule 3. Again, a non-overlapping channel is allocated based on the first cluster and the previous cluster allocated. In Fig. 5, the red sequence numbers show the channel allocation by rule 3.

Figure 6 shows the second step of the channel allocation which allocates non-overlapping channels to one-hop neighbors of the first allocation area.

Figure 7 shows the third step of the channel allocation which allocates non-overlapping channels to one-hop neighbors of the second allocation area. The red sequence numbers show the channel allocation by rule 3.

Figure 8 shows the fourth step of the channel allocation which allocates non-overlapping channels to one-hop neighbors of the third allocation area. The red sequence numbers show the channel allocation by rule 3.

Figure 9 shows that non-overlapping channels are assigned to all clusters in the sensor network. The proposed algorithm successfully allocates exclusive channels between adjacent clusters by using four channels in a very complicate cluster topology.

6 Conclusions

The inter-cluster channel synchronization in the cluster-based sensor networks is important. To solve the channel conflicts of our previous study, we proposed the collision-free channel allocation algorithm.

The execution of the proposed algorithm shows the result of successful non-overlapping channel allocation between adjacent clusters in various and complex cluster topologies. As the future research, the practical protocol and procedures that can be applied in real cluster sensor networks must be studied.

Acknowledgements. This work was funded by the BK21+ program of the National Research Foundation of Korea (NRF). This research was also supported by The Leading Human Resource Training Program of the Regional Neo industry through the National Research Foundation of Korea (NRF) funded by the Ministry of Science, ICT and Future Planning (216C000360).

References

1. Heinzelman, W.R., Chandrakasan, A., Balakrishnan, H.: Energy-efficient communication protocol for wireless microsensor networks. In: Proceedings of the Hwaii International Conference on System Science, pp. 1–10, January 2000
2. Iyengar, S.S., Brooks, R.R.: Distributed Sensor Networks: Sensor Networking and Applications. CRC Press, Boca Raton (2013)
3. Branch, J.W., Giannella, C., Szymanski, B., Wolff, R., Kargupta, H.: In-network outlier detection in wireless sensor networks. J. Knowl. Inf. Syst. **34**(1), 23–54 (2013)
4. Bhanumathi, V., Dhanasekaran, R.: Path discovery and selection for energy efficient routing with transmit power control in MANET. Malays. J. Comput. Sci. **26**(2), 124–139 (2013)
5. Hong, T.-P., Cheng-Hsi, W.: An improved weighted clustering algorithm for determination of application nodes in heterogeneous sensor networks. J. Inf. Hiding Multimed. Signal Process. **2**(2), 173–184 (2011)
6. Karaboga, D., Okdem, S., Ozturk, C.: Cluster based wireless sensor network routing using artificial bee colony algorithm. Wirel. Netw. **18**(7), 847–860 (2012)
7. Lin, H., Uster, H.: Exact and heuristic algorithms for data-gathering cluster-based wireless sensor network design problem. IEEE/ACM Trans. Netw. **22**(3), 903–916 (2014)
8. Baziana, P.A., Pountourakis, I.E.: A channel reuse strategy with adaptive channel allocation for all-optical WDM networks. Opt. Switch. Netw. **10**(3), 246–257 (2013)
9. Karaoglu, B., Heinzelman, W.: Cooperative load balancing and dynamic channel allocation for cluster-based mobile ad hoc networks. IEEE Trans. Mob. Comput. **14**(5), 951–963 (2015)
10. Leem, I.T., Wu, M., Kim, C.: A MAC scheme for avoiding inter-cluster collisions in wireless sensor networks. In: 2010 the 12th International Conference on Advanced Communication Technology (ICACT), pp. 284–288, 7–10 February 2010

11. Wu, M., Leem, I., Jung, J.J., Kim, C.: A resource reuse method in cluster sensor networks in ad hoc networks. In: Pan, J.-S., Chen, S.-M., Nguyen, N.T. (eds.) ACIIDS 2012. LNCS, vol. 7197, pp. 40–50. Springer, Heidelberg (2012). doi:10.1007/978-3-642-28490-8_5

12. Wu, M., Ahn, B., Kim, C.: A channel reuse procedure in clustering sensor networks. In: Applied Mechanics and Materials, vol. 284–287, pp. 1981–1985 (2012)

13. Wu, M., Ha, S., Abdullah, T., Kim, C.: Exclusive channel allocation methods based on four-color theorem in clustering sensor networks. In: Camacho, D., Kim, S.-W., Trawiński, B. (eds.) New Trends in Computational Collective Intelligence Studies in Computational Intelligence, vol. 572, pp. 107–116. Springer, Cham (2015)

14. Allaire, F., Swart, E.R.: A systematic approach to the determination of reducible configurations in the four-color conjecture. J. Comb. Theor. Ser. B **25**, 339–362 (1978)

15. Robertson, N.: The four-colour theorem. J. Comb. Theor. Ser. B **70**, 2–44 (1997)

16. Thomas, R.: An update on the four-color theorem. Not. Am. Math. Soc. **45**, 848–859 (1998)

17. Eliahow, S.: Signed diagonal flips and the four color theorem. Eur. J. Comb. **20**, 641–647 (1999)

18. Hodneland, E., Tai, X.-C., Gerdes, H.-H.: Four-color theorem and level set methods for watershed segmentation. Int. J. Comput. Vis. **82**, 264–283 (2009)

19. Wu, M., Park, H., Zhu, W., Kim, C.: A solution for local channel collisions in sensor networks. Int. J. Control Autom. **9**, 151–162 (2016)

20. Mary, W., Kim, C.: A cost matrix agent for shortest path routing in ad hoc networks. J. Netw. Comput. Appl. **33**, 646–652 (2010)

Feature Selection Techniques for Improving Rare Class Classification in Semiconductor Manufacturing Process

Jae Kwon Kim[1], Kyu Cheol Cho[2], Jong Sik Lee[1], and Young Shin Han[3(✉)]

[1] Department of Computer Science and Information Engineering,
Inha University, Incheon, South Korea
jaekwonkorea@naver.com, jslee@inha.ac.kr
[2] Department of Computer Science, Inha Technical College, Incheon, South Korea
kccho@ingatc.ac.kr
[3] Department of Computer Engineering, Sungkyul University, Anyang, South Korea
hanys@sungkyul.ac.kr

Abstract. In order to enhance the performance, rare class prediction are to need the feature selection method for target class-related feature. Traditional data mining algorithms fail to predict rare class, as the class imbalanced data models are inherently built in favor of the majority of class-common characteristics among data instances. In the present paper, we propose the Euclidean distance- and standard deviation-based feature selection and over-sampling for the fault detection prediction model. We study applying the semiconductor manufacturing process control in fault detection prediction. First, the features calculate the MAV (Mean Absolute Value) median values. Secondly, the MeanEuSTDEV (the mean of Euclidean distance and standard deviation) are used to select the most appropriate features of the classification model. Third, to address the rare class over-fitting problem, oversampling is used. Finally, learning generates the fault detection prediction data-mining model. Furthermore, the prediction model is applied to measure the performance.

Keywords: Semiconductor manufacturing process · Fault detection prediction · Feature selection · Oversampling · MeanEuSTDEV

1 Introduction

Rare class prediction needs the feature selection method, because rare class-related feature in order to enhance the performance. However, conventional data mining methods fail to predict rare class, because most of class- imbalanced data models are inherently built in favor of the majority of class-common characteristics among data instances [1].

The semiconductor manufacturing process is a very complicated process and the structure of the data to be extracted from the process is very complex [2]. Among semiconductor manufacturing processes, some can be detected fault in the FAB process. Therefore, the prediction in the FAB process is important in order to produce the final product. Hence, the pass/fail (regular/irregular) classification technique is necessary in

© ICST Institute for Computer Sciences, Social Informatics and Telecommunications Engineering 2017
J.J. Jung and P. Kim (Eds.): BDTA 2016, LNICST 194, pp. 40–47, 2017.
DOI: 10.1007/978-3-319-58967-1_5

the FAB process; the fault detection prediction before final production can improve quality and reliability [3].

Classification method of data-mining can classify pass/fail using semiconductor's various data. In order to generate the classification model, the data preprocessing, including cleaning, feature selection, oversampling, etc., is crucial [1]. Feature selection can increase accuracy of classifying prediction by eliminating unnecessary attributes, while choosing necessary attributes from high-dimension data set [4]. Extract dataset in the FAB process needs the feature selection method, because the extracted dataset from sensor is very complex.

In the present paper, we propose the Euclidean distance- and standard deviation-based feature selection and over-sampling for the fault detection prediction model. We study applying the semiconductor manufacturing process control in fault detection prediction from the SECOM dataset [5]. First, the features of Semiconductor calculate the MAV (Mean Absolute Value) median values. Secondly, the MeanEuSTDEV (Means of Euclidean distance and standard deviation) are used to select the most appropriate features of the classification model. Third, to address the rare class over-fitting problem, oversampling is used. Finally, learning generates the fault detection prediction data-mining model.

2   Methodology

We built a fault detection prediction model using the SECOM dataset [5]. SECOM dataset is the FAB data collected by 590 sensors from the semiconductor manufacturing process. The SECOM dataset consists of the records of 1,567 samples and 590 features. Among the record of 1,567, the fail class is 104 (encoded as 1), the pass class is 1463 (encoded -1). In order to increase accuracy of the rare class prediction model, we have to choose features among the 590 features related to the target class. The imbalance of the pass and fail classes, in addition to the large number of metrology data obtained from 590 sensors, makes this dataset difficult to accurately analyze.

We mainly focused on devising a feature selection method on data-mining techniques to build an accurate model for rare class detection. The framework of our study is shown in Fig. 1.

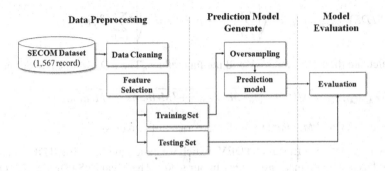

Fig. 1. Frame work

The feature selection method in our study ranges from simply removing a feature with a constant value ('NaN') and missing values (over 60% of values are missing), to statistics-based analysis, such as chi-square, gain ratio, and PCA (Principal Component Analysis). Furthermore, we propose the MeanEuSTDEV to analyze the discrimination of each feature. On the prediction model-building phase, we applied five methods to induce the rare class prediction model, namely LR (Logistics Regression), BPN (Back Propagation Network), SVM (Support Vector Machine), C5.0 (Decision Tree), and KNN (K-nearest neighbor).

Our proposed method for generating a classification model to rare class from the SECOM dataset unfolds in the following steps:

- Data cleaning

First, input the feature data of 590. Remove in the case of a single value. Second, remove the feature in the case of over 60% of NaN (not available) and missing values of record of 1,567.

- Feature selection

We propose the MeanEuSTDEV (Means of Euclidean distance and Standard deviation) based on statistical index. The statistical criterion method can evaluate the distance between two scatter groups (separation index) and directly address the variation of feature in the same group (compactness index). The statistical index should be addressed in both separation and compactness index [6]. ED (Euclidean distance) is the most common use of the distance measure. ED as a separation index. In addition, STDEV is the most robust and widely used measure of variability. STDEV is used as a compactness index. Calculating the MeanEuSTDEV involves the following steps:

(1) Divide data into two class (pass class and fail class)
(2) Calculate the each attribute values using MAV (Mean Absolute Value), median and STDEV (Standard Deviation). MAV and $STDEV$ is defined as

$$MAV_{class(n)} = \frac{1}{n} \sum_{k=1}^{n} |x_k| \tag{1}$$

$$STDEV_{class(n)} = \sigma = \sqrt{\sum_{k=1}^{n} (x_k - m)^2 / n} = \sqrt{(\sum_{k=1}^{n} x_k^2 / n) - m^2} \tag{2}$$

(3) Calculate the ED using MAV and median values. The $ED_{(MAV, Median)}$ is define as

$$ED_{(MAV,Median)} = \sqrt{(MAV_{Pass} - MAV_{Fail})^2 + (Median_{Pass} - Median_{Fail})^2} \tag{3}$$

where, MAV and $Median$ are the feature mean of two class.

(4) The ratio between ED and STDEV, which we called the MeanEuSTDEV index, is used as a statistic measured index in our study. The MeanEuSTDEV index can be expressed as follows:

$$MeanEuSTDEV = ED_{(MAV,Median)}/STDEV_{(Pass,Fail)} \qquad (4)$$

Where $STDEV_{(Pass,Fail)}$ is the average between standard deviation of two classes (pass and fail).

The best performance of classification is obtained when ED is high and the STDEV is low. Hence, MeanEuSTDEV should be large to obtain better performance.

(5) Determine the features that are higher than the average by using descriptive statistics; each value is calculated with MeanEuSTDEV.

- Oversampling

Separate the data into two datasets: the training dataset (70%; total 1,099 records; pass 1,026, fail 73) and the testing dataset (30%; total 468 records; pass 437 fail 31).

Increase the number of records including the fail class in the training data by duplicating the fail class to be same amount as the pass class.

- Prediction model and evaluation

Build a rare class prediction model with LR, ANN, SVM, C5.0, and KNN.

The confusion matrix is used to compare the sensitivity (TP Rate), specificity (TN Rate), precision and accuracy. Confusion matrix is shown in Fig. 2. (TP: True Positive, FP: False Positive, FN: False Negative, TN: True Negative)

		Predict Class	
		Class= -1 (Pass)	Class= 1 (Fail)
Actual Class	Class=-1 (Pass)	TP	FN
	Class= 1 (Fail)	FP	TN

$Sensitivity (TP\ Rate) = TP/(TP + FN)$

$Specificity (TN\ Rate) = TN/(FP + TN)$

$Accuracy = TP + TN/(TP + TN + FP + FN)$

$Precision = TP/(TP + FP)$

Fig. 2. Confusion matrix

3   Experimental

We used JAVA jdk 1.8, Weka, and IBM SPSS Modeler 14.2 for the experiment. To measure the performance of the experiment, we compared the results of chi-square, gain ratio, PCA, and MeanEuSTDEV. Preprocessing results for the fault detection prediction model are shown in Fig. 3.

First, data cleaning is using 309 features by removing 271 features that are over 60% of NaN (not available) and missing values among 590 features. Second, feature selection determines ultimate 117 features among 309 by using the MeanEuSTDEV (average 0.1645). From the feature selection results, feature 104 is the best feature as compared to other features (see Fig. 4). Feature 104 obtains the MeanEuSTDEV of 0.841. Feature 104 is higher than the secondary feature 60 (ca. 0.760). Moreover, as shown in Figs. 5 and 6, feature 162 has the highest ED, but the STDEV of this feature is bad. Hence, the

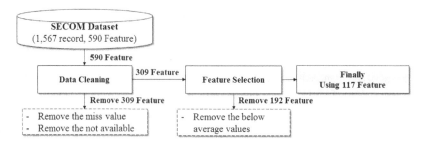

Fig. 3. Preprocessing result

MeanEuSTDEV of feature 162 is not good. However, feature 60 is good group in ED, MeanEuSTDEV. Furthermore, feature 104 is good group in STDEV, MeanEuSTDEV.

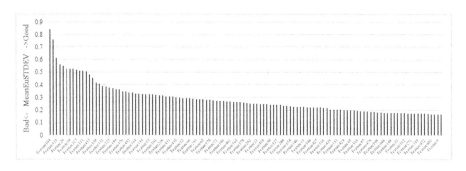

Fig. 4. Bar plot of MeanEuSTDEV of 117 feature

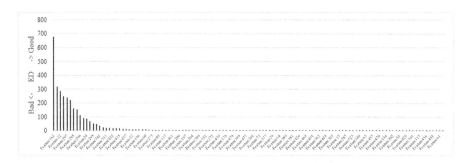

Fig. 5. Bar plot of ED of 117 feature

Next, for generation of the prediction model and experiment, we separated the dataset into the training set, 70% (total 1,099 record; pass 1,026, fail 73) and the testing set, 30% (total 468 record; pass 437 fail 31). Also, the number of training set's fail was set to 953 using oversampling. Finally, the fault detection prediction model was generated using LR, ANN, SVM, C5.0, KNN. The fault detection prediction model's confusion matrix is shown in Table 1. Each model's performance measure is shown in Figs. 7, 8, 9 and 10.

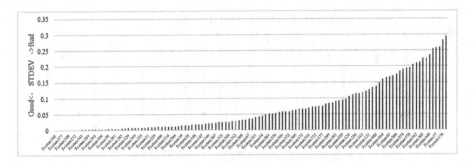

Fig. 6. Bar plot of STDEV of 117 feature

Table 1. Confusion matrix

Model	Feature selection	TP	FN	FP	TN
Chi-Square	LR	317	12	120	19
	ANN	377	23	60	8
	SVM	327	12	110	19
	C 5.0	402	20	35	11
	KNN	387	50	21	10
Gain ratio	LR	319	11	118	20
	ANN	342	14	92	17
	SVM	328	10	109	21
	C 5.0	412	22	25	9
	KNN	391	46	24	7
PCA	LR	303	13	134	18
	ANN	335	14	102	17
	SVM	336	13	101	18
	C 5.0	412	22	25	9
	KNN	391	46	24	7
Non (only cleaning)	LR	357	22	80	9
	ANN	367	23	70	8
	SVM	415	28	22	3
	C 5.0	414	28	23	3
	KNN	390	26	47	5
eanEuSTDEV	LR	353	17	84	14
	ANN	392	26	45	5
	SVM	409	28	28	3
	C 5.0	418	27	19	4
	KNN	386	24	51	7

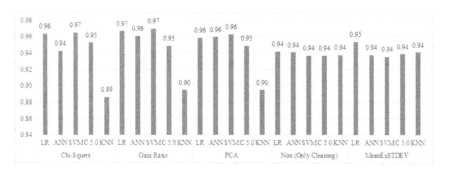

Fig. 7. Performance of sensitivity (TP Rate)

Fig. 8. Performance of specificity (TN Rate)

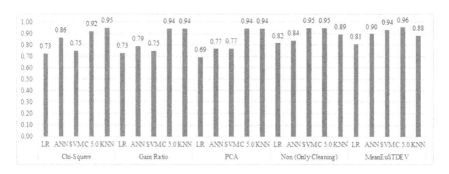

Fig. 9. Performance of precision

The MeanEuSTDEV for C5.0 is highest in precision and accuracy. In other words, the MeanEuSTDEV for C5.0 is useful in feature selection of the semiconductor manufacturing process. Gain ratio is the highest in average of sensitivity (94.8%). PCA is highest in specificity (18.1%) The reason of all model's specificity is lower than 18%, the distribution of pass/fail is imbalance. Therefore, a solution of the data unbalance problem is necessary.

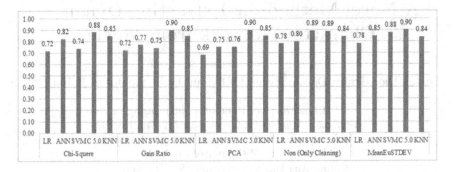

Fig. 10. Performance of accuracy

4   Conclusion

The problem of rare class prediction is important in many real world applications, including the FAB process of semiconductor manufacturing. For a higher accuracy of the rare class prediction model, we need feature selection related to rare class. In this study, we proposed the the Euclidean distance- and standard deviation- based feature selection for the fault detection prediction model. The proposed MeanEuSTDEV calculates the MAV and median value of each feature and extracts the feature using the Euclidean distance and the combination between ED and STDEV. This generates the prediction model using LR, ANN, SVM, C5.0, and KNN. The MeanEuSTDEV for C5.0 is demonstrated to have a better performance than any other feature selection technique.

Acknowledgement. This work was funded by the Ministry of Science, ICT and Future Planning (NRF-2015R1C1A2A01051452).

References

1. Chomboon, K., Kerdprasop, K., Kerdprasop, N.: Rare class discovery techniques for highly imbalance data. In Proceeding International Multi Conference of Engineers and Computer Scientists, vol. 1 (2013)
2. May, G.S., Spanos, C.J.: Fundamentals of Semiconductor Manufacturing and Process Control. Wiley, New York (2006)
3. Purnomo, M.R.A., Dewi, I.H.S.: A manufacturing quality assessment model based-on two stages interval type-2 fuzzy logic. In: IOP Conference Series: Materials Science and Engineering, vol. 105, no. 1, pp. 012044. IOP Publishing (2016)
4. Arif, F., Suryana, N., Hussin, B.: Cascade quality prediction method using multiple PCA+ID3 for multi-stage manufacturing system. IERI Procedia **4**, 201–207 (2013)
5. SEmi COnductor Manufacturing (2010). http://www.causality.inf.ethz.ch/repository.php
6. Phinyomark, A., Hirunviriya, S., Limsakul, C., Phukpattaranont, P.: Evaluation of EMG feature extraction for hand movement recognition based on Euclidean distance and standard deviation. In: International Conference on IEEE (ECTI-CON), pp. 856–860 (2010)

A Novel Method for Extracting Dynamic Character Network from Movie

Quang Dieu Tran[1], Dosam Hwang[1], O.-Joun Lee[2], and Jason J. Jung[2($\boxtimes$)]

[1] Department of Computer Engineering,
Yeungnam University, Gyeongsan 712-749, Korea
dieutq@gmail.com, dosamhwang@gmail.com
[2] Department of Computer Engineering,
Chung-Ang University, Seoul 156-756, Korea
j2jung@gmail.com

Abstract. In this decade, the number of movies is increasing rapidly. Many studies have been proposed to assist users in movie understanding. In which, these methods are taken into account movie content analysis using social network for discovering relationships among characters and so on. However, these methods have shown some unsatisfactorily in dynamic changing of multimedia contents such as the character's relationships over time. For overcoming this issue, we proposed a novel method for extracting dynamic character network from a movie.

Keywords: Dynamic social network · Storytelling analysis · Multimedia analysis

1 Introduction

Today, the number of movies is increasing rapidly. The demand of an effective method for discovering useful information and the story of movie is raising and becoming a challenge task. Various methods and techniques have been proposed for overcoming these issues, in which social network analysis is one of efficient technique for analyzing the storytelling of a movie. In this regards, the story of a movie could be design as a network with a node represents a character and edges represent relationships among characters. The strength of a relationship shows to audiences how a relationship important is. Study from Park et al. showed that we can extract a character network from their's dialogs, and roles of characters are determined based on character network analysis [1]. Moreover, Weng et al. proposed a method for extracting a character network based on the character occurrences [6]. However, such methods have not considered to dynamic changing of character network over time.

Recent years, social network analysis has become a most popular method for discovering useful information from a movie. While using social network for analyzing a movie, an object is represented by a node and edges represent the relationships among them. Regarding this issue, many approaches have been proposed to analyze the content using character network analysis which is extracted

© ICST Institute for Computer Sciences, Social Informatics and Telecommunications Engineering 2017
J.J. Jung and P. Kim (Eds.): BDTA 2016, LNICST 194, pp. 48–53, 2017.
DOI: 10.1007/978-3-319-58967-1_6

from a movie. However, these methods are focused on static social network only [1,6]. In general, a static network is represented as a graph with nodes and edges. This network do not deal with time dimension - one of important factors. In order to address this issue, some approaches have been proposed for analyzing the content of a multimedia document using dynamic social network analysis [2,3]. Exploring movie content is also challenge task. Recent research are focused on extracting a character network based on character's occurrences and co-occurrences. By analyzing this network, the role of characters is discovered as our previous work [4,5]. However, such methods are not considered to timing characteristics of character network, which will show some effective information from movie. This study takes into account proposing a novel method to extract dynamic character network based on character's occurrences and co-occurrences of a movie for overcoming this issue.

This paper is organized as follows. Section 2 discusses a novel method and an algorithm for extracting dynamic character network from a movie. Section 3 describes results and discussion of proposed method. Conclusion and future work of this study are described in Sect. 4.

2   Dynamic Character Network

Movie has a set of characters. These characters play an important role for narrating story based on character's interaction and their occurrences. In general, a character network from the movie is described by a set of nodes and edges where a node represents a character and edges represent the relationships among them. So that, this study takes into account these features for extracting dynamic character network from a given movie.

Regarding the extraction of dynamic network, we applied our previous work for representing the occurrences of characters during movie playback. In this regards, the occurrences of characters are timing characteristics and represent as a set of occurrences sequences, which contain start and end of occurrence time. The sequences of the occurrences of a character is described as the following. Let $\mathcal{C}$ be the set of character in a movie.

Definition 1 (Occurrence Sequence). *The occurrence sequence of a character is defined as the following.*

$$\mathcal{S}_{c_i} = \left\langle \left[t^s_{1_i}, t^e_{1_i} \right], \left[t^s_{2_i}, t^e_{2_i} \right], ..., \left[t^s_{k_i}, t^e_{k_i} \right] \right\rangle \tag{1}$$

where $t^s_{k_i}$ and $t^e_{k_i}$ is the starting and ending time of character c_i who occurs in a movie playback, c_i is a character in $\mathcal{C}$, $i = [1..N_C]$, $k \in N$.

Let c_i and c_j are two characters in a movie. $\mathcal{S}_{c_i}$ and $\mathcal{S}_{c_j}$ are two occurrence sequences of c_i and c_j. We compute the co-occurrences sequence of two characters c_i and c_j as the following.

Definition 2 (Co-Occurrence Sequence). *The co-occurrence sequence of character c_i and character c_j is defined as the following.*

$$\mathcal{A}(c_i, c_j) = \left\langle S_{c_i} \cap S_{c_i} \right\rangle \tag{2}$$

Let $\mathcal{A}_{(c_i,c_j)}$ is the set of co-occurrence sequence of characters c_i and c_j in a movie. We compute total length of character c_i co-occurs to character c_j as the following.

$$l(\mathcal{A}(c_i, c_j)) = \sum_{a_i \in \mathcal{A}(c_i,c_j)} (a_i) \tag{3}$$

Besides, $n(\mathcal{A}_{(c_i,c_j)})$ is the number of co-occurrences between character c_i and c_j.

Regarding dynamic character network analysis, we compute the total of time that characters co-occur and the number of time that them occur at each time t during the playback of a movie.

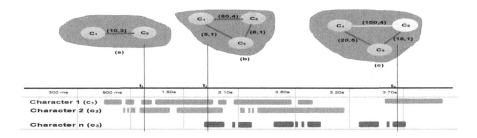

Fig. 1. Dynamic Character Network construction. (a) Character Network at the time t_i; (b) Character Network at the time t_j; (c) Character Network at the time t_v.

In our previous study, we defined CoCharNet as a set of nodes and edges as the following [4]. In which character's occurrences are indexed by "on-screen" visually methods.

Definition 3 (CoCharNet). *CoCharNet is described by a undirected-weight graph as the following*

$$G = \langle C, E \rangle$$

where $C = \{c_1, c_2, ..., c_k\}$ is the set of characters, and E is the set of relationships among them in a movie, in which E is calculated as follows.

$$E = \left\langle (c_i, c_j, l(A_((c_i, c_j)), n(A_{c_i,c_j}) \right\rangle \tag{4}$$

where means character c_i has a relationship with character c_j by total time of co-occurrences $l(A_(c_i, c_j))$ and number of co-occurrences $n(A_{c_i,c_j})$.

Dynamic character network is undirected weighted graph that the topology of this network are changed over time, we define dynamic character network by applying the timing characteristics of the occurrences as the following.

Definition 4 (Dynamic Character Network). *Let L is total length of a movie. Dynamic character network from a movie is described by a sequence of undirected graph as the following.* $\mathcal{D} = \left\langle G_1, G_2, ..., G_L \right\rangle$

where $G_t = \langle C_t, E_t \rangle$ is a undirected weighted graph of character at the time t, C_t is the set of characters at the time t, $0 < t < L$, E_t is the set of relationships among characters in C_t at the time t as follows. The set of relationships among characters C_t is described as the following.

$$E = \left\langle (c_i, c_j, l(A_{(c_i, c_j)})_t, n(A_{c_i, c_j})_t \right\rangle \tag{5}$$

where means character c_i has a relationship with character c_j by total time of co-occurrence $l(A_{c_i, c_j})_t$ and number of co-occurrence $n(A_{c_i, c_j})_t$ at the time t.

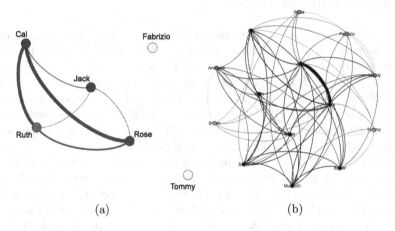

Fig. 2. Dynamic Character Network from Titanic (1997) Movie. (a) t = 30 min; t = 3 h and 10 min.

Figure 1 illustrates the construction of the dynamic character network which is extracted from the given movie. In this regards, at the time t_i, characters c_1 and c_2 are set as active and relationships among them will be calculated. This process is same at the time t_j and t_v, characters c_1 and c_2 and c_3 are set as active depend on their occurrence and co-occurrence. In this regards, relationships among characters will be recalculated at the certain time t.

In order to extract dynamic social network from a movie. Algorithm for extracting dynamic character network is described as the following.

Algorithm 1 illustrates the proposed method for extracting a dynamic character network based on the occurrences and co-occurrences of characters in the given movie.

Algorithm 1. Extracting Dynamic CoCharNet

1: **INPUT** $\mathcal{A}_\mathcal{C}$ is the set of character occurrence intervals;
2: **OUTPUT** Dynamic CoCharNet $\mathcal{D}$;
3: **procedure** DYNAMIC_COCHARNET($\mathcal{D}$)
4:      **for** e **do**ach time t
5:          **while** (existing a_i in $\mathcal{A}_\mathcal{C}$) and ($[t_k^{start}, t] \subseteq \mathcal{A}_\mathcal{C}$) **do**
6:              *compute* $l(\mathcal{A}_\mathcal{C})$;
7:              *compute* $n_{\mathcal{A}_\mathcal{C}}$;
8:              *create nodes appeared in* a_i;
9:              *create an edge in* $\mathcal{G}_t$ *with total time of characters co-occurrence*
10:              $l(\mathcal{A}_\mathcal{C})$ *and number of co-occurrence* $n(\mathcal{A}_{c_i,c_j})$ *at time* t;
11:              $\mathcal{D} = \mathcal{D} + \mathcal{G}_t$
12:          **end while**;
13:      **end for**;
14:      **return** Dynamic CoCharNet $\mathcal{D}$;
15: **end procedure**

3   Results and Discussion

Difference to static network, dynamic network uses node movements or endogenous process over time in the underlying network structure. Figure 2 illustrates the construction of character network from Titanic (1997) movie. At the beginning, *Cal (Caledon Hockley), Ruth (Ruth DeWitt Bukater)* and *Rose (Rose DeWitt Bukater)* are the main characters and they play an important role. But at the end, *Rose (Rose DeWitt Bukater) and Jack (Jack Downson)* are the characters that play important role of the movie. Characters and relationships among them in dynamic character network are changed over time.

In order to help the audiences in movie understanding, dynamic social network analysis is a good way for understanding movies content. By using this network, we can understand how do relationships among characters start and grow during movie playback. Character's occurrences and co-occurrences are able to use for extracting a dynamic character network but other features including character's emotions, activities, and so on should be considered for getting more performance. Next period can be achieved by using this features.

4   Conclusion

Social network analysis has become an important way for analyzing movies content. Many studies have been proposed for extracting social network and analyzing it to discover hidden information from them. However, these methods have considered in static character network for analyzing movie storytelling. In this study, we have proposed a method for extracting dynamic character network based on the occurrences and co-occurrences of characters in a movie. Based on this network, how are relationships among character changing overtime will be described.

Acknowledgments. This work was supported by the National Research Foundation of Korea (NRF) grant funded by the Korea government (MSIP) (NRF-2014R1A2A2A05007154).

References

1. Park, S.-B., Oh, K.-J., Jo, G.-S.: Social network analysis in a movie using character-net. Multimedia Tools Appl. **59**(2), 601–627 (2012)
2. Rand, D.G., Arbesman, S., Christakis, N.A.: Dynamic social networks promote cooperation in experiments with humans. Proc. Natl. Acad. Sci. **108**(48), 19193–19198 (2011)
3. Sarkar, P., Moore, A.W.: Dynamic social network analysis using latent space models. ACM SIGKDD Explor. Newslett. **7**(2), 31–40 (2005)
4. Tran, Q.D., Hwang, D., Lee, O.-J., Jung, J.E.: Exploiting character networks for movie summarization. Multimedia Tools Appl. **76**(8), 10357–10369 (2017). doi:10.1007/s11042-016-3633-6
5. Tran, Q.D., Jung, J.E.: CoCharNet: extracting social networks using character co-occurrence in movies. Univ. Comput. Sci. **21**(6), 796–815 (2015)
6. Weng, C.-Y., Chu, W.-T., Wu, J.-L.: RoleNet: movie analysis from the perspective of social networks. IEEE Trans. Multimedia **11**(2), 256–271 (2009)

Handling Uncertainty in Clustering Art-Exhibition Visiting Styles

Francesco Gullo[1], Giovanni Ponti[2], Andrea Tagarelli[3],
Salvatore Cuomo[4], Pasquale De Michele[4],
and Francesco Piccialli[5($\boxtimes$)]

[1] R&D Department, UniCredit, Via Molfetta 101, 00171 Rome, Italy
gullof@acm.org
[2] DTE-ICT-HPC, ENEA Portici, P.le E. Fermi 1, 80055 Portici, NA, Italy
giovanni.ponti@enea.it
[3] DIMES, University of Calabria, 87036 Rende, CS, Italy
andrea.tagarelli@unical.it
[4] DMA, University of Naples "Federico II",
Via Cupa Cintia 21, 80126 Naples, Italy
{salvatore.cuomo,pasquale.demichele}@unina.it
[5] DIETI, University of Naples "Federico II",
Via Mezzocannone 8, 80100 Naples, Italy
francesco.piccialli@unina.it

Abstract. Uncertainty is one of the most critical aspects that affect the quality of Big Data management and mining methods. Clustering uncertain data has traditionally focused on data coming from location- based services, sensor networks, or error-prone laboratory experiments. In this work we study for the first time the impact of clustering uncertain data on a novel context consisting in visiting styles in an art exhibition. We consider a dataset derived from the interaction of visitors of a museum with a complex Internet of Things (IoT) framework. We model this data as a set of uncertain objects, and cluster them by employing the well-established UK-medoids algorithm. Results show that clustering accuracy is positively impacted when data uncertainty is taken into account.

Keywords: Uncertain objects · Clustering · Data mining · Cultural heritage data

1 Introduction

In the last decade, "Veracity" has been named the fourth "V" referred to the Big Data paradigm in addition to Volume, Velocity and Variety. This attribute emphasizes the importance addressing managing of uncertainty inherent within several types of data. It refers to the level of reliability associated with certain types of data. In this scenario, handling uncertainty in data management requires more and more importance if we consider the wide range of Big Data applications. Some data can be considered inherently uncertain, for example: sentiment in humans; GPS sensors bouncing among the skyscrapers of New York; weather conditions; and clearly the future. The term

© ICST Institute for Computer Sciences, Social Informatics and Telecommunications Engineering 2017
J.J. Jung and P. Kim (Eds.): BDTA 2016, LNICST 194, pp. 54–63, 2017.
DOI: 10.1007/978-3-319-58967-1_7

uncertainty describes anubiquitous status of the information as being produced, transmitted, and acquired in real-world data sources. Exemplary scenarios are related to the use of location-based services for tracking moving objects and sensor networks, which normally produce data whose representation (attributes) is imprecise at a certain degree. Imprecision arises from the presence of noisy factors in the device or trans- mission medium, but also from a high variability in the measurements (e.g., locations of a moving object) that obviously prevents an exact representation at a given time. This is the case virtually for any field in scientific computing, and consequently for a plethora of application fields, including: pattern recognition (e.g., image processing), bioinformatics (e.g., gene expression microarray), computational fluid dynamics and geophysics (e.g., weather forecasting), financial planning (e.g., stock market analysis), GIS applications to distributed network analysis [1].

For data management purposes, uncertainty has been traditionally treated at the attribute level, as this is particularly appealing for inductive learning tasks [22]. In general, attribute-level uncertainty is handled based on a probabilistic representation approach that exploits probability distributions describing the likelihood that any given data tuple appears at each position in a multidimensional domain region; the term *uncertain objects* is commonly used to refer to such data tuples described in terms of probability distributions defined over multidimensional domain regions.

Clustering of uncertain objects has traditionally been employed to categorize data coming from location-based services, sensor networks, or error-prone laboratory experiments. In this work we focus for the first time on studying how handling data uncertainty impacts the performance of clustering methods in a novel context of vis- iting styles in art exhibition. We consider a dataset derived from the analysis of how visitors of a museum interact with mobile devices such as smartphones or tablets. We model this data as a set of uncertain objects, and apply the UK-medoids algorithm [19] to obtain clusters of similar visiting styles. We compare such a visiting-style grouping with a ground truth obtained by a well-established classification methodology, which classifies visiting styles into four categories (ant, butterfly, fish, grasshopper) based on the values of some exemplar parameters, such as the percentage of viewed artworks or the average time spent in interacting with artworks [5, 8–10]. F-measure results confirm the claim that clustering accuracy increases when data uncertainty is taken into account in the process.

The rest of the paper is organized as follows: Sect. 2 describes some preliminaries on clustering techniques of uncertain data and Sect. 3 presents the case study. More- over in Sect. 4 we report some experiments on accuracy and efficiency of K-medoids algorithm applied to our case study. Finally conclusions close the paper.

2 Preliminaries on Clustering of Uncertain Data

Data clustering is a central problem in pattern recognition, knowledge discovery, and data management disciplines. Given a set of objects represented in a multidimensional space, the objective is to infer an organization for these objects into groups, also called *clusters*, according to some notion of affinity or proximity among the objects. Two general desiderata for any clustering algorithm is that each of the discovered clusters should be cohesive (i.e., comprised of objects that are very similar to each other) and that the clusters are well-separated from each other. A major family of clustering algorithms is referred to as partitional clustering [2, 7, 17], whose general scheme is to produce a partitioning of the input set of objects by iteratively refining the assignment of objects to clusters based on the optimization of some criterion function. This approach can be computationally efficient when a proper notion of cluster prototype is defined and used to drive the assignment of objects to clusters. Typically, a cluster prototype is defined as the mean object in the cluster (centroid), or an object that is closest to each of the other objects in the cluster (medoid). K-means [20] and K-medoids [19] are two classic algorithms that exploit the notions of centroid and medoid, respectively.

In this paper we exploit a clustering approach originally designed in the research of uncertain data mining. To this purpose, we can refer to a relatively large corpus of studies developed in the last decade [3, 13–16, 18, 25]. In this work we focus on the uncertain counterpart of K-medoids, named *UK-medoids*, which was proposed in [13]. This algorithm overcomes two main issues of the uncertain K-means (UK-means) [3]: (i) the centroids are regarded as deterministic objects obtained by averaging the expected values of the pdfs of the uncertain objects assigned to a cluster, which may result in loss of information; (ii) the adopted Expected Distance between centroids and uncertain objects requires numerical integral estimations, which are computationally inefficient.

Given a dataset D of uncertain objects and a number k of desired output clusters, the UK-medoids algorithm starts by selecting a set of k initial medoids (uniformly at random or, alternatively, by any ad-hoc strategy for obtaining wellseparated medoids). Then, it iterates through two main steps. In step 1, every object is assigned to the cluster corresponding to the medoid closest to the object. In step 2, all cluster medoids are updated to reflect the object assignments of each cluster. The algorithm terminates when cluster stability is reached (i.e., no relocation of objects has occurred with respect to the previous iteration).

One of the strength point of UK-medoids is that it employs a particularly accurate distance function designed for uncertain objects, which hence overcomes the limitation in accuracy due to a comparison of the expected values of the object pdfs. Also, the uncertain distance for every pair of objects are computed once in the initial stage of the algorithm, and subsequently used at each iteration. The combination of the above two aspects has shown that UK-medoids outperforms UK-means in terms of both effectiveness and efficiency.

3   A Case Study: Styles of Visit in an Art Show

As case study has been considered the art show *"The Beauty and the Truth"*[1]. Here, Neapolitan works of art dating from the late XIX and early XX centuries have been shown. The sculptures have been exposed in the monumental complex of *San*

---

[1] http://www.ilbellooilvero.it

Domenico Maggiore, located in the historical centre of Naples. During the event, we have collected log files related to 253 visitors thanks to an Internet of Things deployed framework [4, 6]. The analysis of their behaviours within the cultural space has enabled us to define a classification of the visiting styles. In the literature, there exist several research papers that focus on this objective.

As a starting point for our classification we have considered the work in [23], where authors have proposed a classification method based on a comparison between behaviours of museum visitors and four "typical" animals (i.e., ant, fish, butterfly and grasshopper). Moreover, we have resorted to the work presented in [24], where, recalling the above mentioned approach, authors have introduced a methodology based on two unsupervised learning approaches for validating empirically their model of visiting styles. Finally, in [5, 8–10], we have proposed a classification technique able to discover how visitors interact with a complex Internet of Thinghs (IoT) framework, redefining the visiting styles' definition. We have considered the behaviours of spectators in connection with the use of the available supporting technology, i.e., smart-phones, tablets and other devices. For completeness, we report a brief description below.

A visitor is considered:

- an *ant* (**A**), if it tends to follow a specific path in the exhibit and intensively enjoys the furnished technology;
- a *butterfly* (**B**), if it does not follow a specific path but rather is guided by the physical orientation of the exhibits and stops frequently to look for more media contents;
- a *fish* (**F**), if it moves around in the center of the room and usually avoids looking at media content details;
- a *grasshopper* (**G**), if it seems to have a specific preference for some preselected artworks and spends a lot of time observing the related media contents.

The four visiting styles are characterized by three different parameters, assuming values in [0, 1]: a_i, τ_i and v_i. More in detail, for the i-th visitor, we denote by:

- a_i, the percentage of viewed artworks;
- τ_i, the average time spent by interacting with the viewed artworks;
- v_i, that measures the quality of the visit, in terms of the sequence of crossed sections (i.e., path).

The classification of the visiting styles is obtained following the scheme summarized in Table 1.

Table 1. Characterization of the visiting styles.

Visiting style	a_i	τ_i	v_i
A	≥ 0.1	Negligible	≥ 0.58
B	≥ 0.1	Negligible	<0.58
F	<0.1	<0.5	Negligible
G	<0.1	≥ 0.5	Negligible

As we can observe, values $a_i \geq 0.1$ characterize both **A**s and **B**s, while values $a_i <$ 0.1 are related to **F**s and **G**s. Moreover, the parameter τ_i does not influence the classification of **A**s and **B**s, while values $\tau_i < 0.5$ are typical for **F**s and values $\tau_i \geq 0.5$ are inherent in **G**s. Finally, the parameter v does not influence the classification of **F**s and **G**s, whereas values $v \geq 0.58$ are related to **A**s and values $v < 0.58$ characterize **B**s. We recall that, each parameter is associated with a numerical value normalized between 0 and 1. The thresholds values $\bar{a} = 0.1$, $\bar{\tau} = 0.5$ and $\bar{v} = 0.58$ have been obtained after a tuning step, in which we have resorted to the K-means clustering algorithm to discover data groups reflecting visitors' behaviours in all the sections of the exhibit. More details, about how these values have been set, are reported in [11].

4   Experimental Evaluation

We devised an experimental evaluation aimed to assess the ability in clustering uncertain objects of the algorithm proposed in [13] and discussed in Sect. 2. We consider the dataset derived from the analysis of how visitors of a museum interact with an IoT framework, according to the methodology described in Sect. 3. We model this data as a set of uncertain objects, and apply the UKmedoids algorithm [19] to obtain clusters of similar visiting styles. The ultimate goal of our evaluation is to compare such a visiting-style grouping with a ground truth obtained by a well-established classification methodology defined [5, 8–10] (described in Sect. 3), and show that our method outperforms a baseline clustering method that does not take uncertainty into consideration.

4.1   Evaluation Methodology

Dataset. Experiments were executed by exploiting the dataset populated with data coming from the above mentioned log files. In the following, we report a description of the dataset resorting, for simplicity of representation, to the ARFF Weka format (see Fig. 1 for more details). Notice that, the dataset is characterized by: (i) 253 objects (i.e., the visitors); (ii) 3 attributes (i.e., a, τ, and v), which reflect, for each visitor, the parameters a_i, τ_i and v_i, described in Sect. 3; (iii) 4 classes (i.e, A, B, F and G), corresponding to the already cited typical animals. Moreover, observe that tuples contain the symbol "?" for some attribute values that are not significant for the classification. In other words, accordingly with the classification rules summarized in Table 1, for **A**s and **B**s we neglect attribute *tau* and for **F**s and **G**s we neglect attribute v.

The selected dataset is originally composed by deterministic values. For this reason, we needed to synthetically generate the uncertainty. Notice that, in order to adapt the dataset to the algorithm in [13], the neglected values have been substituted with the numerical approximation 0.0. In substance, this can be assimilated to a first kind of perturbation.

```
@RELATION ARTWORKS
@ATTRIBUTE a NUMERIC [0..1]
@ATTRIBUTE τ NUMERIC [0..1]
@ATTRIBUTE v NUMERIC [0..1]
@ATTRIBUTE class {A,B,F,G}
@DATA
...
0.0836653386454,0.846588116217,?,G
0.0478087649402,0.317966675258,?,F
0.119521912351,?,0.714285714286,A
0.175298804781,?,0.470303571429,B
...
```

Fig. 1. The dataset in the ARFF Weka format

For the univariate case, we needed to define the region for the interval of uncertainty $I^{(h)}$ and the related pdf $f^{(h)}$ for the region $I^{(h)}$, for all the $a^{(h)}$, $h \in [1..m]$ attributes of the o object. We randomly chose the interval region $I^{(h)}$ as in the subinterval $[min_{oh,}$ $max_{oh}]$, and these two boundaries are the minimum (i.e., $min_{oh)}$ and the maximum (i.e., $max_{oh)}$) deterministic values for h (i.e., the attribute) taking into account the objects that are part of the same ideal classification for o. Regarding $f^{(h)}$, a continuous formulation of the density function has been taken into account, that is *Uniform*, together with a discrete mass function, that is *Binomial*. We properly set the parameters for the Binomial distribution in order to have the mode in correspondence of the original deterministic value of the attribute h-th of the o object.

Clustering Validity Criteria. In order to evaluate quality the clustering in output, we resorted to the availability of the classification originally provided in the dataset. Indeed, following the natural cluster paradigm, the higher the clustering solution is similar to the reference classification, the higher is the quality achieved. *F-measure* [21] is a well known external criterion used to evaluate clustering solution, which exploits *Recall* and *Precision* notions from the Information Retrieval field.

Overall Recall (R) and Precision (P) can be computed by means of a macroaveraging strategy performed on local values as:

$$R = \frac{1}{H} \sum_{i=1}^{H} \max_{j \in [1..K]} R_{i,j}, \qquad P = \frac{1}{H} \sum_{i=1}^{H} \max_{j \in [1..K]} P_{i,j},$$

Overall F-measure is defined as the harmonic mean of P and R as:

$$F = \frac{2PR}{P+R}$$

Settings. The calculation of the distances involves integral computation, and we do it by exploiting the sample list coming from the pdfs. We resorted to a sampling method based on the classical *Monte Carlo*.[2] A tuning phase has been preliminary done in order to set in the proper way the sample number S; the strategy was based on a choice of S producing an accuracy level that another $S' > S$ was not able to improve significantly.

4.2   Results

Accuracy. Accuracy tests have the objective evaluate the impact of dealing with uncertainty in a clustering-based analysis. For this reason, we are interesting in comparing clustering results achieved by our UK-medoids algorithm on the dataset with uncertainty w.r.t. the ones achieved by K-medoids algorithm on the dataset with deterministic values.

In Table 2 we report only results on the univariate model (multivariate model carried out similar results). More in detail, here we highlight the differences, in terms of F-measure percentage gains, between UK-medoids (both binomial and uniform) and deterministic K-medoids. It can be observed that UK-medoids achieves higher accuracy results w.r.t. K-medoids, that are slight for binomial distribution (0.043%), but relevant for uniform one (6.227%). In general, we can notice that introducing uncertainty in the dataset and handling it in the clustering task with our proper UK-medoids algorithm leads to improve the effectiveness of the results.

Table 2. UK-medoids' performance results compared with deterministic K-medoids in terms of F-measure percentage.

pdf	UK-medoids gain
Binomial	0.04298805%
Uniform	6.22712192%

Efficiency. To evaluate the efficiency of UK-medoids, we measured time performances in clustering uncertain objects.[3] Figure 2 shows the total execution times (in milliseconds) obtained by UK-medoids on our dataset. Notice that, we calculated the sum of the times obtained for the pre-computing phase (i.e., uncertain distances computation), together with the algorithm runtimes. Here, it can be noted that by using a uniform pdf we obtain execution times about 2 times faster than those achieved with a binomial pdf. This is due to the fact that a binomial pdf requires to process a higher number of samples w.r.t. a uniform pdf.

---

[2] We used the SSJ library, available at http://www.iro.umontreal.ca/ ∼ simardr/ssj/

[3] Experiments were conducted on an ENEA server of CRESCO4 HPC cluster hosted in Portici [12] – http://www.cresco.enea.it/

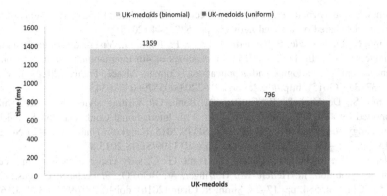

Fig. 2. Clustering time performances

5 Conclusion

In this paper we addressed the topic of how data collected by an IoT system through mobile devices in a cultural environment could be opportunely exploited and analysed. The main goal is to infer useful knowledge about visitors. Real data are generally affected of a large degree of uncertainness and to deal with this drawback, here we propose a clustering approach based on K-medoids algorithms. Nevertheless the limitation of a not very large dataset, first results encourage us to deeply investigate this approach, in order to better analyse data collected from a real cultural heritage scenario. Moreover, with the aim to improve the performance of the proposed method, in future works we will intend to better adapt the uncertain interval and the pdf, defined on this set, to our problem.

References

1. Aggarwal, C.C.: Managing and Mining Uncertain Data: Advances in Database Systems, vol. 35. Kluwer, Boston (2009). http://dx.doi.org/10.1007/978-0-387-09690-2
2. Bello-Orgaz, G., Jung, J., Camacho, D.: Social big data: recent achievements and new challenges. Inf. Fusion **28**, 45–59 (2016)
3. Chau, M., Cheng, R., Kao, B., Ng, J.: Uncertain data mining: an example in clustering location data. In: Proceedings of PAKDD Conference, pp. 199–204 (2006)
4. Chianese, A., Marulli, F., Piccialli, F., Benedusi, P., Jung, J.: An associative engines based approach supporting collaborative analytics in the internet of cultural things. Future Gener. Comput. Syst. **66**, 187–198 (2016)
5. Chianese, A., Piccialli, F.: Improving user experience of cultural environment through IoT: the beauty or the truth case study. Smart Innov. Syst. Technol. **40**, 11–20 (2015)
6. Chianese, A., Piccialli, F., Riccio, G.: Designing a smart multisensor framework based on Beaglebone Black Board. In: Park, J., Stojmenovic, I., Jeong, H., Yi, G. (eds.) Computer Science and its Applications. LNEE, vol. 330, pp. 391–397. Springer, Heidelberg (2015)

7. Cuomo, S., De Michele, P., Galletti, A., Piccialli, F.: A cultural heritage case study of visitor experiences shared on a social network, pp. 539–544 (2015)
8. Cuomo, S., De Michele, P., Galletti, A., Pane, F., Ponti, G.: Visitor dynamics in a cultural heritage scenario. In: DATA 2015 - Proceedings of 4th International Conference on Data Management Technologies and Applications, Colmar, Alsace, France, 20–22 July 2015, pp. 337–343 (2015). http://dx.doi.org/10.5220/0005579603370343
9. Cuomo, S., De Michele, P., Galletti, A., Ponti, G.: Visiting styles in an art exhibition supported by a digital fruition system. In: 11th International Conference on Signal-Image Technology and Internet-Based Systems, SITIS 2015, Bangkok, Thailand, 23–27 November 2015, pp. 775–781 (2015). http://dx.doi.org/10.1109/SITIS.2015.87
10. Cuomo, S., De Michele, P., Galletti, A., Ponti, G.: Classify visitor behaviours in a cultural heritage exhibition. In: Helfert, M., Holzinger, A., Belo, O., Francalanci, C. (eds.) DATA 2015. CCIS, vol. 584, pp. 17–28. Springer, Cham (2016). doi:10.1007/978-3-319-30162-4_2
11. Cuomo, S., De Michele, P., Galletti, A., Ponti, G.: Influence of some parameters on visiting style classification in a cultural heritage case study. In: Pietro, G., Gallo, L., Howlett, R., Jain, L. (eds.) Intelligent Interactive Multimedia Systems and Services 2016. Smart Innovation, Systems and Technologies, vol. 55, pp. 567–576. Springer, Cham (2016). http://dx.doi.org/10.1007/978-3-319-39345-2_50. iIMSS - IS07: Internet of Things: Architecture, Technologies and Applications Invited Session of KES 2016
12. Ponti, G., et al.: The role of medium size facilities in the HPC ecosystem: the case of the new CRESCO4 cluster integrated in the ENEAGRID infrastructure. In: International Conference on High Performance Computing and Simulation, HPCS 2014, Bologna, Italy, 21–25 July 2014, pp. 1030–1033 (2014)
13. Gullo, F., Ponti, G., Tagarelli, A.: Clustering uncertain data via K-medoids. In: Greco, S., Lukasiewicz, T. (eds.) SUM 2008. LNCS, vol. 5291, pp. 229–242. Springer, Heidelberg (2008). doi:10.1007/978-3-540-87993-0_19
14. Gullo, F., Ponti, G., Tagarelli, A.: Minimizing the variance of cluster mixture models for clustering uncertain objects. In: Proceedings of IEEE ICDM Conference, pp. 839–844 (2010)
15. Gullo, F., Ponti, G., Tagarelli, A.: Minimizing the variance of cluster mixture models for clustering uncertain objects. Stat. Anal. Data Min. 6(2), 116–135 (2013)
16. Gullo, F., Tagarelli, A.: Uncertain centroid based partitional clustering of uncertain data. PVLDB 5(7), 610–621 (2012)
17. Jain, A., Dubes, R.: Algorithms for Clustering Data. Prentice-Hall, Upper Saddle River (1988)
18. Jiang, B., Pei, J., Tao, Y., Lin, X.: Clustering uncertain data based on probability distribution similarity. IEEE Trans. Knowl. Data Eng. 25(4), 751–763 (2013)
19. Kaufman, L., Rousseeuw, P.J.: Finding Groups in Data: An Introduction to Cluster Analysis. Wiley, New York (1990)
20. MacQueen, J.B.: Some methods for classification and analysis of multivariate observations. In: Proceedings of Berkeley Symposium on Mathematical Statistics and Probability, pp. 281–297 (1967)
21. van Rijsbergen, C.J.: Information Retrieval. Butterworths, London (1979)
22. Sarma, A.D., Benjelloun, O., Halevy, A.Y., Nabar, S.U., Widom, J.: Representing uncertain data: models, properties, and algorithms. VLDB J. 18(5), 989–1019 (2009). doi:10.1007/s00778-009-0147-0

23. Veron, E., Levasseur, M., Barbier-Bouvet, J.: Ethnographie de l'exposition. Paris, Biblioth`eque Publique d'Information, Centre Georges Pompidou (1983)
24. Zancanaro, M., Kuflik, T., Boger, Z., Goren-Bar, D., Goldwasser, D.: Analyzing museum visitors' behavior patterns. In: Proceedings of 11th International Conference on User Modeling 2007, UM 2007, Corfu, Greece, 25–29 June 2007, pp. 238–246 (2007)
25. Züfle, A., Emrich, T., Schmid, K.A., Mamoulis, N., Zimek, A., Renz, M.: Representative clustering of uncertain data. In: Proceedings of KDD Conference, pp. 243–252 (2014)

Using Geotagged Resources on Social Media for Cultural Tourism: A Case Study on Cultural Heritage Tourism

Tuong Tri Nguyen[1], Dosam Hwang[2], and Jason J. Jung[3(✉)]

[1] Department of Computer Science, Hue University of Education, Hue, Vietnam
tuongtringuyen@gmail.com
[2] Department of Computer Engineering,
Yeungnam University, Gyeongsan, Korea
dosamhwang@gmail.com
[3] School of Computer Engineering, Chung-Ang University, Seoul, Korea
j2jung@gmail.com

Abstract. In recently, the smart tourism applications are raising the scale of data to an unprecedented level. A new emerging trend in social media namely to collect and introduce cultural heritage by geotagged resources were being focused on. The paper aims to deliver a way to collect geotagged cultural heritage resources from social networking services by using the keyword and user's position (GPS signal) to deliver smart interactions between visitors in a smart tourism environments. A large number of the cultural heritage information repositories are explored by using the user's geo-location. Therefore, from determining a user's position and context, the data that are related to cultural heritages nearby that location is collected such as photos, tags, comments. In the next step, the system is implemented for classifying and filtering the collected data belongs to users interest (e.g., the ancient capital, citadel, dynasty, tomb); determining the representative photo and important tags of each place; recommending the famous places based on photo distribution and users criteria to tourists. The experimental results show the map based on criteria given by users that contained useful information to visit some cultural heritages mentioned.

Keywords: Smart cultural tourism · Cultural heritage · Social media data · Geotags resources

1 Introduction

Nowadays, the users can exchange information easily based on the social networking services [1, 2]. Additionally, there is more and more sharing information associated with geographic locations [3]. Smart tourism applications are used popularity in recently [4–6] in which people is equipped with mobile devices can interact with cultural objects, sharing and producing data. Furthermore, they can also require useful personalized services to enhance the quality of their cultural experience [7].

In this paper, a large number of general and cultural heritage information repositories are explored based on the user's position. Therefore, from determining a user's

© ICST Institute for Computer Sciences, Social Informatics and Telecommunications Engineering 2017
J.J. Jung and P. Kim (Eds.): BDTA 2016, LNICST 194, pp. 64–72, 2017.
DOI: 10.1007/978-3-319-58967-1_8

position and context, the data that are related to cultural heritages nearby that location is collected such as photos, tags, comments. The application is implemented based on geotagged resources from social network services including Flickr and Instagram. It analyzes these resources to introduce to people as a smart way for guidance them during their trip.

This research aims to present a new emerging trend in social media namely introducing and collecting cultural heritage based on geotagged resources. Herein, we introduce a smart cultural tourism (SCT) system by using geotagged resources on social networking services. SCT is used to collect and analyze geotagged data from different resources on social media by using the keyword and user's position to deliver smart interactions between visitors in a smart tourism environments.

The paper is organized as follows. Section 2 refers to some studies related to smart cultural tourism and cultural tourism heritage by using resources on social networking services. In Sect. 3, we state the problem and describe some basic knowledge related to the smart tourism and location-aware based on geotagged cultural heritage resources. For the Sect. 4, we introduce the methodology for exploiting the smart cultural tourism based on geotagged cultural heritage on social networking services. Section 5 shows the experimental results that was conducted to evaluate the proposed method. Section 6 draws some conclusion and states the future works.

2 Related Works

There are many studies using social networking services as a tool to develop their tourism to introduce and to exhibit the cultural heritage to tourism in recently [7–11]. A smart context-aware system to model the context evolution, adopting a graph structure is presented by the authors [7], named Context Evolution System. The authors have showed an example of context evolution modeling inside an art exhibition named The Beauty or The Truth, located in Naples within the monumental complex of San Domenico Maggiore, Italy.

The authors in [8] have presented a platform that used social media as a data source to support the decisions of policymakers in the context of smart tourism destinations initiatives. Besides, they have showed that it is possible to identify the nationality, language of the posts, points of agglomeration and concentration of visitors based on analyzing 7.5 million tweets. Their results can be applicable to the effective management of smart tourism destinations.

Through network analysis, the authors [9] investigated utilization of Facebook by local Korean governments for developing the tourism services. They also indicated that Korean local governments Facebook pages are related to the Facebook system as part of a smart tourism ecosystem.

Besides that a framework of summarizing tourism information in response to popular tourist locations were introduced by the authors [10]. They have crawled the huge travel blog data and applied for a frequent pattern mining method in order to detect efficient travel routes in the identified popular locations. Moreover, the system had also provided a travel information to a Chinese online tourism service company.

In one of other research, the authors [11] introduce their recommendation systems and adaptive systems. The system have been introduced in travel applications to support the travelers in the decision-making processes.

In this research, we focus on exploiting a large number of geographic tags of cultural heritage resources and incorporating with media data to find a set of attractive photos and useful tags to provide and introduce to tourists some useful information related to their trip.

3   Problem Definition

3.1   Problem

To provide some useful information to tourists, the research problem is defined by two questions as follows. Does the smart cultural tourism actively use geotagged resources to provide tourism information to users for developing tourism? How might the smart cultural tourism be exploited by using social media such as Flickr, Instagram, Face-Book to open cultural heritage tourism widely?

3.2   System Overview

The workflow of SCT is showed in Fig. 1. Its components are described as follows.

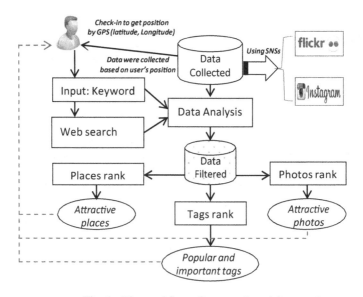

Fig. 1. The workflow of proposed model

SCT aims to uncover the areas and challenges in the execution of smart tourism systems based on geotagged resources from social media for cultural heritage tourism.

In particular, this study will unveil the four key components, principally, 'collecting data related to heritage', 'analyzing data', 'processing data', and 'showing information'.

In order to answer the research questions, we proposed some steps to implement the system for smart cultural tourism by using geotagged resources as follows:

Step 1.   *Collect data based on the position of user;*

Step 2.   *Filter the collected data belongs to users interest (cultural heritage);*

Step 3.   *Determine the representative photo and important tags of each place;*

Step 4.   *Recommend the famous places based on photo distribution and users criteria to visit;*

Step 5.   *Show the map based on criteria given*

4   Solution

4.1   Computation Distribution of Resources

In order to determine the distance between position of photos, our system had used Haversine formula [12]. It is quite popular and frequently used formula when developing a Geographic Information System application or analyzing path and fields. For any two points on a sphere, the Haversine of central angle between them is given by

$$\mathcal{H}\left(\frac{d}{r}\right) = \mathcal{H}(\varphi_2 - \varphi_1) + \cos(\varphi_1)\cos(\varphi_2)\mathcal{H}(\lambda_2 - \lambda_1) \tag{1}$$

where $\mathcal{H}$ is the haversine function $\mathcal{H}(\theta) = \sin^2\left(\frac{\theta}{2}\right) = \frac{1-\cos(\theta)}{2}$

d is the distance between the two points (along a great circle of the sphere; see spherical distance); r is the radius of the sphere; φ_1, φ_2: latitude of point 1 and latitude of point 2, in radians; λ_1, λ_2: longitude of point 1 and longitude of point 2, in radians.

On the left side of Eq. 1, $\frac{d}{r}$ is the central angle, assuming angles are measured in radians (note that φ and λ; can be converted from radians to degrees by multiplying by $\frac{180}{pi}$ as usual).

Solve for d by applying the inverse haversine (if available) or by using the arcsine (inverse sine) function:

$$d = r\mathcal{H}^{-1}(h) = 2r\arcsin(\sqrt{h}) \tag{2}$$

where h is $\mathcal{H}\left(\frac{d}{r}\right)$, or more explicitly:

$$d = 2r\arcsin\left(\sqrt{\mathcal{H}(\varphi_2 - \varphi_1) + \cos(\varphi_1)\cos(\varphi_2)\mathcal{H}(\lambda_2 - \lambda_1)}\right)$$
$$= 2r\arcsin\left(\sqrt{\sin^2\left(\frac{\varphi_2-\varphi_1}{2}\right) + \cos(\varphi_1)\cos(\varphi_2)\sin^2\left(\frac{\lambda_2-\lambda_1}{2}\right)}\right)$$

4.2   Proposed Algorithm

In this section, the distance measure between any two points to cluster photos on the dataset is used. We consider using photos such as a point on the map. They will be clustered based on its position through the distance between points which are described by latitude and longitude and a threshold.

Algorithm 1 Photos clustering

1: Input: a set of photo $x_i \in P$, $n = |P|$, x_i contains *lat, lon*;
2: Output: a set of photo $x_i \in P$ clustered;
3: **procedure** Cluster(P)
4:       $c \Leftarrow 1$ // the number of cluster;
5:       $Cluster(x_1) \leftarrow c$;
6:       **for** $i \in [2, \ldots, n]$ **do**;
7:             $dmin \Leftarrow r$ // the radius of the sphere;
8:             **for** $j \in [1, \ldots, i-1]$ **do**;
9:                   $dist \leftarrow Distance(x_i, x_j)$;
10:                  **if** $dist < dmin$ **then**
11:                        $dmin = dist$;
12:                        $Cluster\ of\ x_i \leftarrow Cluster\ of\ x_j$;
13:                  **end if**;
14:            **end for**;
15:            **if** $dmin > threshold$ **then**
16:                  $c \Leftarrow c + 1$;
17:                  $Cluster\ of\ x_i \leftarrow c$;
18:            **end if**;
19:      **end for**;
20:      Return P; 21:
end procedure;

Additionally, the system used a filtered algorithm based on the similarity between the features of each visiting places and the set of photo's tags in order to remove some photos which are not close to the places corresponding to the clusters. Finally, we had applied the ranking algorithm in [13] with the data filtered by some conditions to show the results.

5   Experiment

5.1   Dataset

The system works base on the user's position to collect data from social networking services. Therefore, to implement with real-world data, we assume that users visit some places around the cultural heritage such as Hue, Hoi An, and My Son in Vietnam. Using the position of these places, the application automatically collected a set of geotagged photos from Flickr and Instagram, the dataset is described in Table 1.

Table 1. Dataset

Cultural heritage	Position	Radius	#Photos	Country
The ancient capital of Hue	16.46, 107.57	10	23557	Vietnam
Hoi An ancient town	15.88, 108.336	10	26144	Vietnam
My Son sanctuary	15.79, 108.107	2	1966	Vietnam

5.2   Experimental Results

Firstly, we collect data from user's position and then get another data based on a keyword that is given by the user. Both of two datasets are used to extract the features of each place. Secondly, the system was implemented to cluster the data to find some interested places and also select the useful tags and attractive photos to represent these places.

The Fig. 2 showed some important tags which are selected based on two factors, tags collecting from geotagged cultural heritage resources from Hue[1] and websites related to the cultural heritage mentioned by keyword given. We had combined them thanks to term frequency of tags and simulate by using TagCrowd[2] application.

Figure 3 introduces the distribution of photo before and after filtering data. Figure 4 shows some attractive photos belong to each places after the user giving some keywords including "heritage, citadel".

Fig. 2. Tags from the cultural heritages in Hue

---

[1] A city of the central in Vietnam.

[2] http://tagcrowd.com.

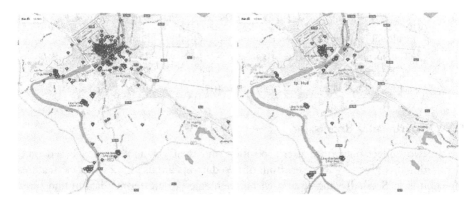

Fig. 3. Map distribution of photo based on user's position (Before and After filtering)

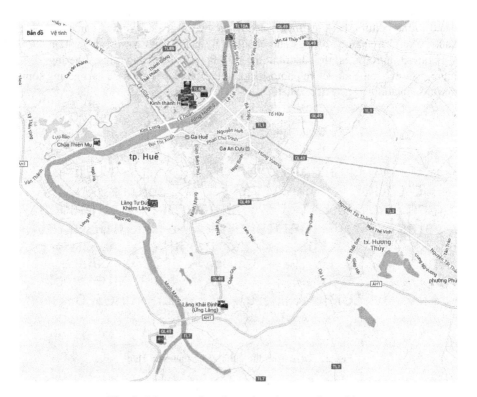

Fig. 4. Map attractive photos based on user's position

6 Conclusion and Future Work

6.1 Conclusion

In this study, we had exploited geotagged resources from social networking services for the purposes of smart cultural tourism development.

The research results indicated that the combination of the semantic tags and media data such as image data bring out many useful information for users during their trip.

Thank to these results, we can foresee that social networking services are likely to be further integrated into the smart cultural tourism in Vietnam and other countries which have many cultural structures tourism.

6.2 Limitations

Using information from social networking services are facing with some difficulties in security such as some photos in Flickr; collecting a list of friends of a user on Instagram, are not allowed because of the privacy of these applications and some user's security mechanisms. These lead to some useful information will not be collected to implement in our system. Therefore, the system is limited by data collected.

6.3 Future Work

In the near future, the consideration of solving big data is one of the goal to research. Because of the size of data is increasing, especially with data on social networking services.

Further, the incorporation of IoT sensors is becoming a new trend. The system will be integrated functions of mobile, sensors to collect information and interactions between different users in the mobile space tourist.

Acknowledgment. This work was supported by the National Research Foundation of Korea (NRF) grant funded by the Korea government (MSIP) (NRF-2014R1A2A2A05007154). Also, this work was supported by the BK21 + program of the National Research Foundation (NRF) of Korea.

References

1. Nguyen, T.T., Hwang, D., Jung, J.J.: Social tagging analytics for processing unlabeled resources: a case study on non-geotagged photos. In: Proceedings of the 8th International Symposium on Intelligent Distributed Computing VIII, IDC 2014, Madrid, Spain, 3–5 September 2014, pp. 357–367 (2014)
2. Nguyen, T.T., Nguyen, H.L., Hwang, D., Jung, J.J.: Pagerank-based approach on ranking social events: a case study with flickr. In: 2015 2nd National Foundation for Science and Technology Development Conference on Information and Computer Science (NICS), pp. 147–152. IEEE (2015)

3. Pham, X.H., Nguyen, T.T., Jung, J.J., Hwang, D.: Extending HITS algorithm for- ranking locations by using geotagged resources. In: Proceedings of 6th International Conference on Computational Collective Intelligence. Technologies and Applications, ICCCI 2014, Seoul, Korea, 24–26 September 2014, pp. 332–341 (2014)
4. Constantinidis, D.: Crowdsourcing culture: challenges to change. In: Borowiecki, K.J., Forbes, N., Fresa, A. (eds.) Cultural Heritage in a Changing World, pp. 215–234. Springer, Cham (2016). doi:10.1007/978-3-319-29544-2_13
5. Gretzel, U., Sigala, M., Xiang, Z., Koo, C.: Smart tourism: foundations and developments. Electron. Markets **25**(3), 179–188 (2015)
6. Tenerelli, P., Demšar, U., Luque, S.: Crowdsourcing indicators for cultural ecosystem services: a geographically weighted approach for mountain landscapes. Ecol. Ind. **64**, 237–248 (2016)
7. Chianese, A., Piccialli, F.: A smart system to manage the context evolution in the cultural heritage domain. Comput. Electr. Eng. **55**, 24–26 (2016)
8. Cacho, A., et al.: Social smart destination: a platform to analyze user generated content in smart tourism destinations. In: Rocha, Á., Correia, A., Adeli, H., Reis, L., Mendonça Teixeira, M. (eds.) New Advances in Information Systems and Technologies. Advances in Intelligent Systems and Computing, vol. 444, pp. 817–826. Springer, Cham (2016)
9. Park, J.H., Lee, C., Yoo, C., Nam, Y.: An analysis of the utilization of facebook by local Korean governments for tourism development and the network of smart tourism ecosystem. Int. J. Inf. Manage. **36**(6), 1320–1327 (2016)
10. Yuan, H., Xu, H., Qian, Y., Li, Y.: Make your travel smarter: summarizing urban tourism information from massive blog data. Int. J. Inf. Manage. **36**(6), 1306–1319 (2016)
11. Etaati, L., Sundaram, D.: Adaptive tourist recommendation system: conceptual frameworks and implementations. Vietnam J. Comput. Sci. **2**(2), 95–107 (2015)
12. Glen, V.B.: Heavenly Mathematics: The Forgotten Art of Spherical Trigonometry. Princeton University Press, Princeton (2013)
13. Nguyen, T.T., Jung, J.J.: Exploiting geotagged resources to spatial ranking by extending HITS algorithm. Comput. Sci. Inf. Syst. **12**(1), 185–201 (2015)

Archaeological Site Image Content Retrieval and Automated Generating Image Descriptions with Neural Network

Sathit Prasomphan[✉]

Department of Computer and Information Science, Faculty of Applied Science,
King Mongkut's University of Technology North Bangkok,
1518 Pracharat 1 Road, Wongsawang, Bangsue, Bangkok 10800, Thailand
ssp.kmutnb@gmail.com

Abstract. This research presents a novel algorithms for generating descriptions of stupa image such as stupa era, stupa architecture by using key points generated from SIFT algorithms and learning stupa description from the generated key points with artificial neural network. Neural network was used for being the classifier for generating the description. We have presented a new approach to feature extraction based on analysis of key points and descriptors of an image. The experimental results for stupa image content generator was analyze by using the classification results of the proposed algorithms to classify era and architecture of the tested stupa image. To test the performance of the purposed algorithms, images from the well-known historical area in Thailand were used which are image dataset in Phra Nakhon Si Ayutta province, Sukhothai province and Bangkok. The confusion matrix of the proposed algorithms gives the accuracy 80.67%, 79.35% and 82.47% in Ayutthaya era, Sukhothai era and Rattanakosin era. Results show that the proposed technique can efficiently find the correct descriptions compared to using the traditional method.

Keywords: Image content retrieval · Neural network · SIFT algorithms · Feature extraction

1 Introduction

Thailand or Siam is one of a country that has a long history in the Southeast Asia. Several temples, palaces, or residences had been developed in each era. However, in present, because of the long time of that place some importance things are broken. Some parts of that place still remain which is interesting to the new young generation who interested in an archaeological site. One of archaeological site which is most importance for studying is stupa. Several architecture of the stupa was created. If we know the stupa architecture, more details of that stupa can be described.

To study the stupa architecture, the shape of each stupa will be considered by finding the shape similarity between the target image and all of images in the database. The purpose of these algorithms is to get information from the interested image. These algorithms also known as image content retrieval algorithms. Nowadays, there are several techniques for getting information from an image [4–6]. For example, a machine learning algorithm was

© ICST Institute for Computer Sciences, Social Informatics and Telecommunications Engineering 2017
J.J. Jung and P. Kim (Eds.): BDTA 2016, LNICST 194, pp. 73–82, 2017.
DOI: 10.1007/978-3-319-58967-1_9

applied such as recursive neural network [2] or convolution neural network [3] which introduced by Richard Socher et al. [2]. In this technique the relevance between image and the sentence was used. Another rule which used the recurrent neural network was presented by Andrej Karpathy and Li Fei-Fei [1]. The language and image relationship was studied. Different sources of image such as Flickr8K, Flickr30K and MSCOCO were used in the research. The problem of using these categories is the complicated of model and input attribute to be used to find the relationship between language structure and image. Another group of model was suggested by using the scoring method to find the relationship between sentence space, image space, and meaning space. This technique was suggested by Ali Farhadi et al. [8]. However, there is some confusion to create the relationship between all of these spaces.

From the limitation of these techniques, the new algorithm for generating image description is developed. The archaeological site image especially stupa image was studied. The research aims to find the architecture of the stupa image which can be used this architecture to generate other descriptions. For example, era of the stupa, when the stupa was builds or where the stupa location frequently occurs. In this research, the combination between SIFT algorithms and neural network was studied. SIFT algorithms was used for generating key points of an image. Neural network was used for classifying the architecture of stupa image. The input attributes of neural network comes from key point.

The remaining of this paper is organized as follows. At first, we show the stupa characteristics. Next, the combination between SIFT algorithms and neural network is discussed in Sect. 2. Results and discussion are discussed in Sect. 3. Finally, the conclusions of the research are presented in Sect. 4.

2   Proposed Algorithms

In this section, the characteristics and examples of stupa image occurred in Thailand were described. The SIFT algorithms for generating key points was detailed. The neural network for classifying architecture of stupa was shown. Finally, the process for generating description to the stupa image was explained.

2.1   Characteristics of Stupa

The stupa or known as chedi is a Buddhist memorial monument usually housing holy relics associated with the Buddha. The stupa shape was created from the shape of an ancient Indian burial mound [7]. Several styles of stupa in Thailand occurred. The architecture's category of stupa in Thailand can be divided into three styles. These categories are divided based on the period of that architecture [7]. However, the overlapped architecture between periods also occurs. The stupa architecture in the Sukhothai period is grouped into the lotus blossom style, the bell-shaped style, the Prang style, etc. The stupa architecture in the Ayutthaya period is classified into the bell-shaped style, the Prang style, etc. Finally, The stupa architecture in can be classified into these categories: the square wooden chedi style, the Prang style, etc. The example of stupa architecture in each period (Sukhothai period, Ayutthaya period, and Rattanakosin period) are shown in Figs. 1, 2, and 3 in orderly.

Fig. 1. Example of stupa in the Sukhothai era (a) the lotus blossom style (b) the bell-shaped style (c) the Prang style (d) the Chomhar style [7].

Fig. 2. Example of stupa in Ayutthaya era [7].

Fig. 3. Example of stupa in Rattanakosin era [7].

2.2 Scale Invariant Feature Transforms (SIFT) Algorithms

Scale-invariant feature transform (SIFT) [9, 10] is an algorithm in computer vision to detect and describe local features in images. The algorithm was published by David Lowe in 1999 [9]. SIFT [9, 10] is an algorithm for calculating the interesting point (key point) in an image and determining the features of key point. The characteristics of these features are invariant to image scale and rotation. They are robust to addition of noise, distortion or brightness. SIFT technique has an advantage that it is not based on a scale or the orientation angle position, which can be used to compare the features more easily and accurately, more precisely. Key point generally refers to a pixel in the image which is changed the orientation from two-dimensional of brightness levels surrounded a key point. The algorithm for findings SIFT key point in the picture is as follows. Firstly, detect key point from the input image (key point detection). In this step, we get a series of x, y coordinates of a key point which is used to provide a description of the key point. For the next step, part of the explanation of key point (key point description) is calculated in form of a vector explain (descriptor vector), which is calculated from the brightness of the pixels in the area surround key point. These vectors are used to describe the series of identity when it appears in photos. After the key point is generated, the process of learning or learning phase is performed. It is the process for matching between the most two similar images. Following are the major stages of computation used to generate the set of image features:

1. **Scale-space extrema detection:** The process for detect the most importance feature of image which not depends on the size or orientation of an image. The process is done by burring image with Gaussian function in each octave. In each octave consists of several burring scale by burring normal scale and increasing scale parameter which effect to the burring image. It is done with octave which in each step reduce the size to half of old octave.
2. **Key point localization:** From the previous process each scale space image will be used to find the key point of image by using the octave with the following equation. The process for getting key point is local maxima/minima in DoG image and finds sub pixel maxima/minima. In the previous process, it will get many key points. The process for reduce number of key points is removing low contrast feature and removing edges.
3. **Orientation assignment:** After specify the key points; it will calculate magnitude and orientation of gradient surrounded the key points for becoming descriptors of key points. After that, this magnitude and gradient of pixel surround key points will be used for generating histogram which x is degree and y is gradient.
4. **Key point descriptor:** Create 16×16 windows and divided to 4×4 window with 16 sets which in each to calculate magnitude and gradient and create histogram with 8 bins. After finished this process the feature vector will be 128 and will be used to the next processes.

The process in each step can be shown in Figs. 4, 5 and 6.

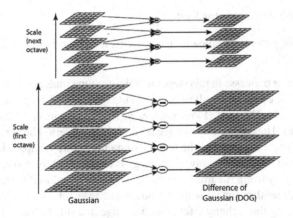

Fig. 4. Difference of Gaussian is obtained as the difference of Gaussian blurring of an image with two different σ, let it be σ and $k\sigma$ [9].

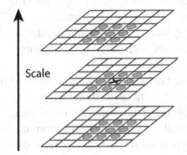

Fig. 5. Once this DoG are found, images are searched for local extrema over scale and space. One pixel in an image is compared with its 8 neighbours as well as 9 pixels in next scale and 9 pixels in previous scales. If it is a local extrema, it is a potential keypoint. It basically means that keypoint is best represented in that scale [9].

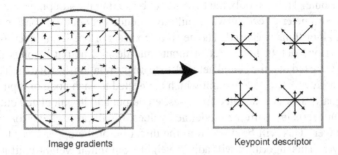

Image gradients   Keypoint descriptor

Fig. 6. After specify the key points; magnitude and orientation of gradient surrounded the key points for becoming descriptors of key points will be calculated [9].

2.3   Image Content Generator System

The processes for generating stupa image contents as shown in Fig. 7 are described in the following section.

1. **Pre-processing process:** In this step, the training stupa images and the testing stupa image is collected in which all of training stupa image come from the well-known historical area in Thailand (Phra Nakhon Si Ayutta province, Sukhothai province and Bangkok). The testing stupa image is taken from camera. The similarity between the training stupa images and the testing stupa image are compared. The transformation from RGB image to gray scale image is necessary. Moreover, to improve the image quality by using the image enhancement is required in this step.

2. **Laplacian algorithm for edge detection:** In this research, the Laplacian algorithm is used for being the technique for detecting edge in a stupa image. The different of intensity of nearest points is measured. Finding the line surrounding the object inside an image is required. We use this process for finding the edge that pass through or near to the interested point. In low quality image, the different of intensity of nearest point is low which effect to the process of finding edge. The foreground and background of brightness may be not covering all of image. In this case the edge may be blurred compared to using the high different of intensity. The information from edge detection is used for the next process.

3. **Feature extraction:** The key point generating from the SIFT algorithm is the main input features in this research. In each stupa image, the vector of key point is used. The key point can be identified the property in each image. Magnitude and orientation of gradient surrounded the key points for becoming descriptors of key points is calculated. 128 attributes from the 128 key point's descriptor are generated. Our assumption in this step is the most two similar images will have the same key point.

4. **Stupa architecture classification:** In this paper, neural networks have been created in order to distinguish for key points of an image. Use the input attribute from the key points and descriptors generated from SIFT algorithms for becoming the input to neural network. We use 128 attributes from the 128 key point's descriptor to be input attribute. Train on only the training set by setting the stopping criteria and the network parameters. Feed-forward multilayer neural network with back propagation learning algorithms was used. The network consists of one input layer, one hidden layer, and one output layer. Set of inputs and desired output belongs to training patterns were fed into the neural network to learn the relationship of data. The process in hidden layer is to adjust weight which connected to each node of input. The root mean square error is calculated from desired output and its calculated output. If the error is not satisfied with the predefined values, it will propagate error back to the former layer. This will be done from the direction of the upper layer towards the input layer. This algorithm will adjust weight from initial weight until it gives the satisfied mean square error. We use this technique for matching the architecture of the stupa between the testing stupa image and all of stupa reference images in database. Inside the stupa reference images database, it will contain stupa image in several architectures, the descriptions of the reference image, stupa architecture,

stupa era and other importance details of stupa. The algorithm will select the most matching architecture of testing stupa image and the reference stupa images.

5. **Stupa description generating process:** After the matching process finished, there is the stupa description generating process. The algorithms will use image descriptions inside the database to show and set description to the testing image.

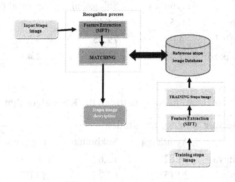

Fig. 7. The process of image content generator system.

3 Experimental Results and Discussions

3.1 Experimental Setup

The stupa image datasets in this research came from the historical area at Phra Nakhon Si Ayutta province, Sukhothai province and Bangkok in which the most stupa architecture in that area is Ayutthaya period, Sukhothai period and Rattanakosin period in ordering. The constraint in each image must have least 80% of image coverage. The brightness of image will be effect to the classification results. So, the pre-processing of image was required. The number of image using in the research was shown in Table 1.

Table 1. Number of samples in the experiments

Era	Number of image
Ayutthaya	67
Sukhothai .	44
Rattanakosin	46
All	**157**

3.2 Performance Indexed

3.2.1 Confusion Matrix

Tables 2, 3 and 4 are the confusion matrices, which are widely used graphical tools that reflect the performance of an algorithm. Each row of the matrix represents the instances of a predicted class, while each column represents the instances of an original class. Thus, it is easy to visualize the classifier's errors while trying to accurately predict each original class' instances. Percentage of the test data being classified to the original stupa image was shown inside the table.

Table 2. Confusion matrix of the proposed algorithms by using sift with neural network

Architecture	Image recognition		
	Ayutthaya	Sukhothai	Rattanakosin
Ayutthaya	**80.67**	4.43	7.35
Sukhothai	4.07	**79.35**	3.29
Rattanakosin	7.43	3.79	**82.47**

Table 3. Confusion matrix of the KNN algorithm

Architecture	Image recognition		
	Ayutthaya	Sukhothai	Rattanakosin
Ayutthaya	**60.82**	12.69	26.49
Sukhothai	20.15	**62.78**	17.07
Rattanakosin	15.67	21.02	**63.27**

Table 4. Confusion matrix of euclidean distance

Architecture	Image recognition		
	Ayutthaya	Sukhothai	Rattanakosin
Ayutthaya	**70.28**	12.88	16.84
Sukhothai	13.58	**75.22**	11.20
Rattanakosin	17.57	8.79	**73.64**

3.2.2   Comparing Algorithms

In this paper, Euclidean distance, SIFT algorithms with neural network, and k-nearest neighbors were used for comparing. We compared each of this method with the proposed algorithms which used to predict the era of that architecture for example Ayutthaya era, Sukhothai era and Rattanakosin era. We used the combination of SIFT algorithms and neural network algorithms.

4   Experimental Results and Discussions

4.1   Experimental Results

To show the accuracy of the proposed stupa image content generator, the stupa architecture classification results were analyzed. Euclidean distance, SIFT algorithms with neural network, and k-nearest neighbors were used for comparing. The classification accuracy was shown in Tables 2, 3 and 4. Table 2 showed the confusion matrix of the proposed algorithms which was the combination between SIFT algorithms and neural network. It gives the accuracy 80.67%, 79.35% and 82.47% in Ayutthaya era, Sukhothai era and Rattanakosin era. Table 3 showed the confusion matrix of using k-nn algorithms. This algorithm gives the accuracy 60.82%, 62.78% and 63.27% in Ayutthaya era, Sukhothai era and Rattanakosin era. Table 4 showed the confusion matrix of using the Euclidean distance. This algorithm gives the accuracy 70.28%, 75.22%, and 73.64% in Ayutthaya era, Sukhothai era and Rattanakosin era. The experimental results confirms that using key points of image which is generated from the SIFT algorithms and using the neural network for training the key points to get the period of a stupa architecture inside an image can be successfully used. Also this predicted architecture's era can be used for generating the description of image as shown in table. The proposed algorithms give the accuracy about 80–85% in average.

Example of using SIFT algorithms for generating key point to match the architecture of stupa with different comparing algorithms for example, Euclidean distance, neural network, and k-nn algorithm can be shown in Table 5.

Table 5. Examples of using sift algorithms for generating key point.

Algorithms	Example	
	Rattanakosin era	Ayutthaya era
SIFT with Euclidean distance		
SIFT algorithms with k-nearest neighbors		
SIFT algorithms with neural network		

5   Conclusions

This research proposes a new algorithm for getting image descriptions via key point descriptor from SIFT algorithms. The key point descriptors of an image are used to distinguish identity of an image. The important feature was extracted from a stupa image. A basic architecture of neural network to reduce the difficulty of the classification process is used. Number of features to be sent to neural network is reduced. The algorithm was tested with a stupa image getting from the real world in historical area in Ayutthaya province, Sukhothai province and Bangkok. The confusion matrix of the proposed algorithms gives the accuracy 80.67%, 79.35% and 82.47% in

Ayutthaya era, Sukhothai era and Rattanakosin era. Results show that the proposed technique can efficiently find the correct descriptions compared to using the traditional method.

Acknowledgment. This research was funded by King Mongkut's University of Technology North Bangkok. Contract no. KMUTNB-59-GEN-048.

References

1. Karpathy, A., Fei-Fei, L.: Deep visual-semantic alignments for generating image descriptions. In: The 2015 IEEE Conference on Computer Vision and Pattern Recognition (CVPR), pp. 1–14 (2015)
2. Socher, R., Karpathy, A., Le, Q.V., Manning, C.D., Ng, A.Y.: Grounded compositional semantics for finding and describing images with sentences. TACL **2**, 207–218 (2014)
3. Zaremba, W., Sutskever, I., Vinyals, O.: Recurrent neural network regularization. arXiv preprint arXiv:1409.2329 (2014)
4. Young, P., Lai, A., Hodosh, M., Hockenmaier, J.: From image descriptions to visual denotations: new similarity metrics for semantic inference over event descriptions. TACL **2**, 67–78 (2014)
5. Hodosh, M., Young, P., Hockenmaier, J.: Framing image description as a ranking task: data, models and evaluation metrics. J. Artif. Intell. Res. **47**, 853–899 (2013)
6. Su, H., Wang, F., Yi, L., Guibas, L.J.: 3D-assisted image feature synthesis for novel views of an object, CoRR http://arxiv.org/abs/1412.0003 (2014)
7. Charuwan, C.: Buddhist Arts of Thailand, Buddha Dharma Education Association Inc., Tullera (1981)
8. Farhadi, A., Hejrati, M., Sadeghi, M.A., Young, P., Rashtchian, C., Hockenmaier, J., Forsyth, D.: Every picture tells a story: generating sentences from images. In: Daniilidis, K., Maragos, P., Paragios, N. (eds.) ECCV 2010. LNCS, vol. 6314, pp. 15–29: Springer, Heidelberg (2010). doi:10.1007/978-3-642-15561-1_2
9. Lowe, D.G.: Object recognition from local scale-invariant features. In: Proceedings of 7th International Conference on Computer Vision (ICCV 1999), Corfu, Greece, pp. 1150–1157 (1999)
10. Lowe, D.G.: Distinctive image features from scale-invariant key points. Int. J. Comput. Vis. **60**(2), 91–110 (2004)

Correcting Misspelled Words in Twitter Text

Jeongin Kim[1], Eunji Lee[1], Taekeun Hong[2], and Pankoo Kim[1(✉)]

[1] Department of Computer Engineering, Chosun University, Gwangju, Republic of Korea
jungingim@gmail.com, eunbesu@gmail.com, pkkim@chosun.ac.kr
[2] Department of Software Convergence Engineering, Chosun University,
Gwangju, Republic of Korea
goodfax2000@naver.com

Abstract. The SNS became popularized by computer, mobile devices, and tablets that are accessible to the Internet. Among SNS, Twitter posts the words of short texts and, it shares information. Twitter texts are the optimal data to extract new information, but as it may contain the information within the limited number of words, there are various limitations. To improve accuracy of extracting information within Twitter texts, the process of calibrating misspelled words shall be taken in advance. In conventional studies to correct the misspelled words of Twitter texts, the relationship between misspelled words and correcting words was resolved by concerning the dependency of co-occurrence words with misspelled words within sentences and morphophonemic similarity, but since the frequency of co-occurrence words of misspelled words is not concerned, it has not resolved to correct misspelled words completely. In this paper, to correct misspelled words in Twitter texts, the use of the character n-gram method concerning spelling information and the word n-gram method concerning frequency of co-occurrence words are to be proposed.

Keywords: Twitter text · Misspelled word · Correcting misspelled words · Character n-gram · Word n-gram

1 Introduction

SNS has become more popularized due to the rapid growth of the use of devices like tablet that is accessible to the Internet. SNS users post their profiles and contents, share messages, pictures, and links, and maintain social relationship through these activities [1]. Among SNS, "Twitter" is one of the most widely used micro blogs. Users transmit a short message of 140 characters called Tweet to share personal opinions and information and follow other users to receive their Tweets [2]. As many users of Twitter may share tweets, it is very influential to propagate information [3]. The emergency landing of passenger flight on Hudson River of New York in 2009 could be taken care of quickly as one of Twitter users announced the accident through Twitter. The texts by Twitter user at the accident were spread faster than conventional news. From the terror of Boston Marathon in 2013, the influence of Tweet could be found. After the terror, Twitter users sold various souvenirs and T-shirts to help victims and hash-tagged "Boston Strong" on their tweets. As this hash-tag "Boston Strong" was spread widely, this became the slogan

© ICST Institute for Computer Sciences, Social Informatics and Telecommunications Engineering 2017
J.J. Jung and P. Kim (Eds.): BDTA 2016, LNICST 194, pp. 83–90, 2017.
DOI: 10.1007/978-3-319-58967-1_10

of strong spirit of Boston against adversity. Furthermore, in the terror of France in 2015, the Twitter text information was used. After the terror in France, a group of Twitter accounts tweeted the phrase of welcoming the attack and also tweeted the phrase of warning about more terrors in the future. Through the tweets by terror group, the group of terror could be predicted. As shown above, the Twitter user may recognize the requested information from the bulk Twitter texts, but in case of machine, to analyze the bulk text information, it is required to have a learning process about the characteristics of documents. Yet, in case of short text like Twitter, since the lexical variants are included to compose sentences to hold the information within the limited number of characters, it is limited to extract the information with the conventional method. Therefore, this study is to propose the solution using character n-gram method for spelling information to correct misspelled words and word n-gram method for frequency of co-occurrence words in Twitter. The paper is composed as follows. In the Sect. 2, the relative studies are explained. In the Sect. 3, the basis of selecting Twitter as the subject of correction of misspelled words is stated. In the Sect. 3.2, the overall system composition and the correcting of misspelled words and evaluation using character n-gram, and word n-gram are stated. In the Sect. 4, the conclusion and the future study are stated to conclude the paper.

2   Related Work

Richard Beaufort [6] has proposed the normalization by sharing the similarity of spelling check method and machine translation method. The normalization of system was based on the training model using word phrases. The validity of the method for French was checked by using the 10-fold cross verification. Choudhury [7] studied about the characteristics of natural language and texts that can be summarized and established the word level model. The Hidden Markov Model was composed to determine the type of word deformed from a normal word and the probability of deforming. The structure of Hidden Markov Model may find the words of types that could not be seen before through the linguistic analysis of SMS data. The variables of Hidden Markov Model may estimate the word arrangement of SMS texts and Standard English parallel phrases through the lesson of machine. Hassan [10] proposes the text normalization system of social media for social media text. Upon the unlabeled text phrase, the n-gram sequence is composed to use the random work similar to the context. By using this method by Hassan, there is no limit in domain and language, and in processing the social media text and in pre-processing stage of NLP application program, it is useful. Kobus [11] proposed the method of automatic normalization by using the uniqueness of text written by a machine like e-mail, blog, and chatting. For French SMS messages, it has normalized spelling by using another method that is not used for the automatic voice recognition device. Liu [12] proposed the method of normalizing the non-standard lexical and many abbreviations used in SMS and Twitter. The approach of character-level block as a divided word was proposed, and it has been combined with the conventional method.

3   Correcting Method and Test Evaluation of Misspelled Word of Tweet Text

This chapter proposes the correction system of misspelled words using character n-gram and word n-gram for misspelled words, the limitation in extraction of information from Twitter. Figure 1 is an overall diagram of system to extract and correct misspelled words from Twitter text.

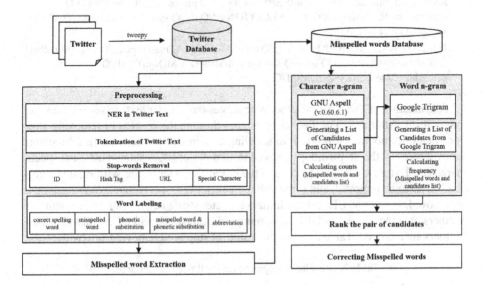

Fig. 1. Overall diagram of system to extract and correct misspelled words from Twitter text.

3.1   Extraction Method of Misspelled Words Within Twitter Text

To collect Twitter texts, the tweepy, the python openAPI, is used. The collected Twitter text is composed of various languages. Since this paper is about the misspelled words from Twitter texts in English, only the Twitter texts in English were extracted to be established as the database of Twitter. The texts stored in the Twitter database goes through the pre-process step. The pre-process step is conducted as the following order: 1. NER in Twitter Text, 2. Tokenization of Twitter Text, 3. Stop-words Removal, 4. Word Labeling. In a process of NER (Named Entity Recognition) in Twitter Text, the Named Entity of organization, location, and person within Twitter texts are NER. Since NER are unnecessary words for correction of misspelled words, they are removed from Twitter texts [17]. The NER used for Twitter texts was the Named Entity Recognition and the Standford NER Software developed at Stanford University. Table 1 shows the example of recognizing the Named Entity from Twitter texts.

Table 1. Example of recognizing the Named Entity from Twitter texts.

Twitter	
1	Hey/O @zaynmalik/O I/O just/O recieved/O my/O order/O fulfillment/O for/O MindOfMine/O thanks/O brother/O
2	I/O added/O a/O video/O 2/O a/O @YouTube/O laylist/Ohttp://youtu.be/ 2gGYWCLwYOI?a/O 50/O Cent/O x/O **Chris/PERSON** Brown/PERSON -/O I/O am/O The/O Man/O -LRB-/O Live/O in/O **Oakland/LOCATION**-RRB-/O
3	Just/O to/O refresh/O **Samantha/PERSON**,/O the/O planes/O hit/O the/O twin/O towers/O and/O **Pentagon/ORGANIZATION** b4/O the/O military/O action/O./O./O #auspol/O #lookitup/O
4	#DLL/O day/O 2tmrw/O!/O Planning/O time/O for/O the/O marketplace/O among/O other/ O fun/O things/O./O CU/O soon/O @CampbellMira/O @MrNgoTDSB/O @teaching24seven/O @ZeliaMCT/O

Tokenization of Twitter Text process separates the Twitter texts based on the space among them. After that, The Twitter texts goes through the process of removing stopword that does not take a big role in showing contents of the context. In a Stopword removal process, the Stopword is removed in an order of ID, hashtag, URL, and special characters. To remove ID and hashtag, if the initial character begins with @, it is recognized as ID to be removed, and if the initial character is #, it is recognized as hashtag to be removed. In case of URL, if the initial character string begins with http, youtu, bit, or others, it is recognized as URL to be removed. At last, the special characters !, $, %, or others are removed. Table 2 shows the results of stopword removal from twitter texts.

Table 2. Results of stopword removal from twitter texts.

Twitter	
1	Hey, I, just, recieved, my, order, fulfillment, for, thanks, brother
2	I, added, a, video, 2, a, playlist, 50, Cent, x, I, am, The, Man, Live, in
3	Just, to, refresh, the, planes, hit, the, twin, towers, and, b4, the, military, action
4	day, 2tmrw, Planning, time, for, the, marketplace, among, other, fun, things, CU, soon

In the Word Labeling process, the Twitter text after removal through NER and tokenization are labeled into five types of correct spelling words (csw), misspelled words (mw), phonetic substitution (ps), misspelled words & phonetic substitution (mp), and abbreviation (ab). The correct spelling words and misspelled words were labeled by using the GNU Aspell dictionary (v6.06), and the chat word dictionary was used to label the phonetic substitution, misspelled words & phonetic substitution, and abbreviation. Table 3 shows an example of word labeling using the chat word dictionary and the GNU Aspell dictionary.

To correct the misspelled words in the Twitter texts, the words labeled as mw among 5 types of labeling (csw, mw, ps, mp, and ab) are extracted. In this study, the misspelled words are corrected by using character n-gram and word n-gram methods

Table 3. Example of word labeling using the chat word dictionary and the GNU Aspell dictionary.

Twitter	
1	(Hey, csw), (I, csw), (just, csw), **(recieved, mw)**, (my, csw), (order, csw), (fulfillment, csw), (for, csw), (thanks, csw), (brother, csw)
2	(I, csw), (added, csw), (a, csw), (video, csw), **(2, ps)**, (a, csw), (playlist, csw), **(50, ps)**, (Cent, csw), (x, csw), (I, csw), (am, csw), (The, csw), (Man, csw), (Live, csw), (in, csw)
3	(Just, csw), (to, csw), (refresh, csw), (the, csw), (planes, csw), (hit, csw), (the, csw), (twin, csw), (towers, csw), (and, csw), **(b4, mp)**, (the, csw), (military, csw), (action, csw)
4	(day, csw), **(2tmrw, ab)**, (Planning, csw), (time, csw), (for, csw), (the, csw), (marketplace, csw), (among, csw), (other, csw), (fun, csw), (things, csw), **(CU, ps)**, (soon, csw)

3.2 Method of Generating Word Pairs to Correct Misspelled Word

In this section, the process of generating a word pair of misspelled word and correct spelling word is be stated. In a process of character n-gram, the misspelled words and the GNU Aspell dictionary are used to generate the list of candidates. To evaluate the similarity in form between misspelled word and candidates, the LCS (Longest common Subsequence) method is used. The LCS algorithm finds the longest common subsequence from two strings of characters. Here, the partial string is different from a substring. The partial string refers to the derived string that may erase some characters but does not change the order. In other words, the partial character string shall be continuous, but the partial string does not have to be continuous. The partial character string of phrase is always a partial string, but the partial string does not have to be partial character string at all-time [18]. In this paper, the length of LCS between misspelled word and candidates are measured. For example, the longest common partial character

Table 4. Length of LCS and the frequency of 3-gram.

Misspelled word	Candidate word	LCS length	3-gram
Recieved	Received	7	223,448
	Relieved	7	55
	Receives	6	0
	Receive	6	3,392
	Revived	6	150
	Receiver	6	0
	Reserved	6	886
	Deceived	6	0
	Receded	6	0
	Recited	6	203
	Relived	6	116
	Perceived	6	80
	Recede	5	0
	Rived	5	0

string is "abegceb", and thus the length of LCS is 7. In a process of word n-gram, the 3-gram of misspelled words is generated, and the frequency of google 3-gram is calculated. The 3-gram of misspelled word is composed of misspelled word and surrounding words. For example, in a sentence, "Hey I just recieved my order fulfillment for thanks brother.", the 3-gram of misspelled word "recieved" can be generated as (I just *), (just * my), or (* my order). For the empty space (*) of 3-gram of misspelled word, the words of candidates composed through character n-gram are entered. The 3-gram of misspelled word is searched on the google 3-gram to calculate the frequency. To generate a word pair of misspelled word and correct spelling word, the length of LCS measured by character n-gram and the 3-gram frequency of misspelled word are used. Table 4 shows the length of LCS between misspelled word and candidates and the frequency of 3-gram of misspelled word.

However, the range of data value by the frequency of 3-gram of misspelled word and the length of LCS are different from each other. Therefore, there shall be a process of normalization to make the ranges of two data to be the same. The Eq. 1 is used for normalization of the length of LCS, and the Eq. 2 is used for normalization of the frequency of 3-gram of misspelled word.

$$NLCS_{length} = \frac{LCS_{length}}{MaxLCS_{length}} \tag{1}$$

$$NF_{3gram} = \frac{f(trigram, googletrigram)}{Maxf(trigram, googletrigram)} \tag{2}$$

At last, to correct misspelled words, a pair of word composed of misspelled word and correcting spelling word is made. To select the correct spelling word of misspelled word, the maximum value of candidates is measured. To measure this maximum value, the sum of length of normalized LCS and frequency of 3-gram of normalized misspelled word is used. The Eq. 3 is used to calculate the sum of length of normalized LCS and frequency of 3-gram of normalized misspelled word.

$$NLCS_{length} + NF_{3gram} \tag{3}$$

Among candidates of misspelled words, the candidate with the maximum value is selected for a pair of misspelled word and correct spelling word. The misspelled word within the Twitter text is substituted with the correcting spelling word of pair of misspelled word under the appearing order of misspelled word.

4   Conclusion

In this paper, the method of correcting misspelled words within the Twitter texts is proposed. The Twitter texts are collected by using tweepy, or python openAPI. The extraction of misspelled words within Twitter texts are done through NER in Twitter Text, Tokenization of Twitter Text, Stop-words Removal, and Word Labeling. A pair of misspelled words extracted is generated by using character n-gram and word n-gram methods. In a character n-gram method, the list of candidates is generated, and the length of LCS is measured. The list of candidates is generated by using the GNU Aspell dictionary. The length of LCS is measured by using the LCS algorithm. In a word n-gram method, the 3-gram is generated with misspelled words and surrounding words, and the frequency of 3-gram is measured. Two words on the left and two words on the right with misspelled words within the Twitter texts are used to be generated. The frequency of 3-gram is measured by searching the google 3-gram. Since the LCS length of candidates measured in character n-gram and word n-gram methods and the frequency of 3-gram have different ranges from each other, and thus they are normalized to be summed up. Among candidates, the word with the maximum value is selected as the correct spelling word to be paired up with a misspelled word. The misspelled words of Twitter texts are corrected by substituting the correct spelling word of pair of misspelled word generated by the method proposed in this paper. As a result of correcting misspelled words within the Twitter texts, the method proposed in this paper is effective.

Acknowledgments. This research was supported by the Human Resource Training Program for Regional Innovation and Creativity through the Ministry of Education and National Research Foundation of Korea (NRF-2014H1C1A1073115) and This research was supported by SW Master's course of hiring contract Program grant funded by the Ministry of Science, ICT and Future Planning (H0116-16-1013).

References

1. Wilson, C., Boe, B., Sala, A., Puttaswamy, K.P.N., Zhao, B.Y.: User intereactions in social networks and their implications. In: Proceedings of the 4th ACM European Conference on Compter Systems, pp. 205–218 (2009)
2. Kim, J., Ko, B., Jeong, H., Kim, P.: A method for extracting topics in news twitter. Int. J. Softw. Eng. Appl. **7**(2), 1–6 (2013)
3. Vespignani, A.: Modelling dynamical processes in complex socio-technical systems. Nat. Phys. **8**, 32–39 (2012)
4. Beaufort, R., Roekhaut, S., Cougnon, L.A., Fairon, C.: A hybrid rule/model-based finite-state framework for normalizing SMS messages. In: Proceedings of the 48th Annual Meeting of the ACL (ACL 2010), pp. 770–779 (2010)
5. Choudhury, M., Saraf, R., Jain, V., Mukherjee, A., Sarkar, S., Basu, A.: Investigation and modeling of the structure of texting language. Int. J. Doc. Anal. Recogn. **10**(3), 157–174 (2007)
6. Hassan, H., Menezes, A.: Social text normalization using contextual graph random walks. In: The 51st Annual Meeting of the Association for Computational Linguistics (ACL 2013), pp. 1577–1586 (2013)
7. Kobus, C., Yvon, F., Damnati, G.: Normalizing SMS: are two metaphors better than one? In: The 22nd International Conference on Computational Linguistics (COLING 2008), pp. 441–448 (2008)
8. Chen, Y.: Improving text normalization using character-blocks based models and system combination. In: The 24th International Conference on Computational Linguistics (COLING 2012), pp. 1587–1602 (2012)

9. Jung, J.J.: Online named entity recognition method for microtexts in social networking services: a case study of twitter. Expert Syst. Appl. **39**(9), 8066–8070 (2012)
10. Longest common subsequence problem. http://en.wikipedia.org/wiki/Longest_common_subsequence_problem

Theoretical Concept of Inverse Kinematic Models to Determine Valid Work Areas Using Target Coordinates from NC-Programs

A Model Comparison to Extend a System-in-Progress as Systems Engineering Task

Jens Weber[(⊠)]

Heinz Nixdorf Institute, Fuerstenallee 11, Paderborn, Germany
jens.weber@hni.upb.de

Abstract. In order to determine valid production results in the area of NC-based tooling machine, complex simulation tools are used. The challenge is the calculation of the material removal, tool paths as well as valid setup positions of workpiece and periphery which leads mostly to high computational time. The descent of the computational effort includes a high portion of systems engineering. This contribution shows a theoretical concept to substitute complex simulation models by calculation models which observes the inverse kinematic behavior of the machine in combination with a NC-parser which estimates valid workpiece positions. The contribution compares model approaches of inverse kinematic problems considering 5 axis tooling machine to determine valid setup positions and minimize theoretical calculation effort.

Keywords: Inverse kinematic · NC-program · NC-command · Quaternion · Translation vector · Rotation matrix

1 Introduction

In order to realize a simulation-based optimization process to determine valid setup position for workpieces during tooling operations, the idea arose to combine meta-heuristics as optimization component and a simulation model of a tooling machine as evaluation component. The simulation model, which is a CAD-based machine model which is controlled by a real control unit leads to high evaluation effort when each potential solution candidates includes a single simulation. The problem of this approach is to shrink the evaluation process in order to reduce the number of single simulation runs of production processes.

The NC-program offers NC-commands which contains the target coordinates to reach the target geometry of the raw material (workpiece). A given software program, so called NC-parser, identifies the tool paths and calculate the trajectories of the program cycles and return the production time depending on zero-point of workpiece-position coordinates. The combination of the NC-parser as fitness component and PSO algorithm as optimization component is a rapid combination to reach a huge number of

© ICST Institute for Computer Sciences, Social Informatics and Telecommunications Engineering 2017
J.J. Jung and P. Kim (Eds.): BDTA 2016, LNICST 194, pp. 91–101, 2017.
DOI: 10.1007/978-3-319-58967-1_11

potential solution candidates. This allows a rapid pre-processing method without using complex simulations. However the system requires a further component which examine unintentional collisions between workpiece, workpiece clamps or machine periphery which is normally given by the complex simulation model. For this problem, there are created several solution concepts and the theoretical model concept of this contribution constitutes a further one to focus the given problem: Development and comparison of inverse kinematic approaches to estimate valid workpiece positions in the machine area. Section two gives an overview about the related work of the basic research area and research project. Section three contains an overview about the system and shows potential inverse kinematic models. Section four offers a statement about the calculation effort of the model approaches and discuss the potential of the models using several role model scenarios to review the approaches. Section five closes the contribution with a conclusion and an outlook.

2   Research Project and Related Work of the Research Project

In order to improve the work preparation process, the research project "*InVorMa*" arose, which is supported by the German Ministry of Education and Research. The goal of the project is to develop a service platform in order to optimize work preparation processes and the identification of optimized and valid setup of production parameters in the area of cutting machining processes. The project contains subprojects to present several solutions to guarantee an optimal job-schedule, suitable machine selection depending on workplace volume as well as setup optimization of production parameters and the distribution to machine instances and computer resources.

In this area, there arose several approaches to decrease the simulation effort, computing time as well as developing an automatically experimental design to identify the best setup position coordinates of workpiece on the machine table which leads to minimal production time by simultaneous collision free positioning of workpieces.

In order to provide more efficiency optimization component, the basis particle swarm Optimization algorithm (PSO) is tested as asynchronous extension to handle stochastic node failures and asynchronous particle evaluation processes in order to shrink the total process time [1]. The contributions [2] illustrates the zero point optimization for workpieces using the metaheuristic PSO as optimization component and the NC-Parser as given software program estimating tool paths and production durations. This contribution acts as proof of concept that the PSO algorithm is correctly configured in order to search the workpiece-zero-point position using machine geometry (3- and 5-axis tooling machines) and real NC-programs in order to minimize tool paths [2]. In order to concern the problem of high simulation effort as well as testing several cluster algorithms in combination with the PSO-NC-parser-concept, the contributions [3–5] offers a conceptual solution.

3   Concept of the System and Inverse Kinematic Models

3.1   System Concept

To identify automatically potential setup parameter for correct production processes, the optimization component generates parameters which represent e.g. workpiece-positions, zero-points, and tool changes. This information are summarized as input data to define a simulation job of a potential production and machine scenario. Each of the potential solutions are pre-evaluated by a NC-parser, which offers the resulting cutting time as well as secondary machine time. In a further loop, the results which lead to the minimal production time are clustered by the K-Means algorithm and each cluster is assigned to a virtual tooling machine (computer resource) by the simulation scheduler. Because of the high simulation effort caused by virtual tooling models which use a real control unit powered by named companies, the evaluation time increase depending on computer resources. From academicals perspective that would not be a high challenge to build a scheduling system to offer an adequate solution. However for practical use in a standard production environment in companies, this procedure is unpractical, so that the idea arose to extend the pre-processing optimization containing optimization algorithm (PSO) and NC-parser and offer an inverse kinematic model along the lines of a real used tooling machine, which represents always a kinematic chain. In this way, invalid work areas are determined before the simulation sessions and jobs are organized to save work preparation time in order to decrease the total number of simulation runs with complex machine models. There is now the challenge to identify the most practical method to model the inverse kinematic for a work preparation platform. The developed architecture is shown in Fig. 1.

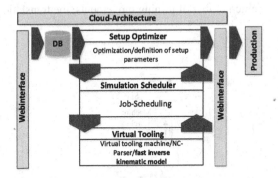

Fig. 1. Schematic system overview

3.2   Concept of the Inverse Kinematic Models and Linked Research Issues

For the solution of the inverse kinematic problems, there are general approaches which are often discussed in literature (see [6, 9–11]): A geometrical solution approach, algebraically solution approach and the numerical solution approach. An often used method especially for robot control is the quaternions algebra. The usefulness of the inverse kinematic model is the calculation of Cartesian coordinate systems in node-coordinates systems [6, 7] which could lead to the problem that the number of potential solution candidates is infinite or no solutions are possible. Figure 2 underlines the most important differences between inverse kinematic and the opposite forward kinematic (see [8, 9]). Figure 2 represents the required in- and output variables.

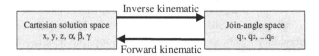

Fig. 2. Schematic overview about the differences between inverse and forward kinematic

The solution space depends on the based technical system and the design of the kinematic structure. For example, the aspired efforts to find a reasonable solution, the inverse kinematic problem has to be comply specific limit values which are given by the physical circumstances of the tooling machine design as kinematic chain. In this contribution, a 5-axis milling machine is the role model for the inverse kinematic model. Because of the kinematic chain of a tooling machine, which is shown in Fig. 3, the inverse kinematic model is built in order to use the combination of the approaches of rotation matrix and translation vector as use case 1. Use case 2 consists of the combination of quaternion and translation vector. In the quaternion algebraic approach translationally movements are feasible because of the screw theory which would lead to a high calculation effort in this context. The Screw Theory means that the turning movements caused by quaternions calculation operations are arranged such a screw thread and after a specific number of turning movements, the translationally distance is feasible (see [11]).

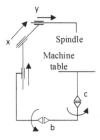

Fig. 3. Schematic kinematic model of a tooling machine

Research Issues which are investigated using the use cases:

1. *Is the combination of several model approaches usable to design the behavior of the inverse kinematic for the role model tooling machine?*
2. *Are the input and output data set required for the inverse kinematic model clear manageable for system engineering?*

3.3 Model for Use Case 1: Combination of Rotation Matrix and Translation Vectors

The degree of freedom of the model for the use cases (see Fig. 3) is assumed as $f = 5$. The toolpaths in the direction of x, y and z represents the coordinate system KS_0. The support coordinate systems KS_1 and KS_2 are placed including the point of origin in the center of the machine table (see Fig. 3). Let the y_{KS_1}-axis be on one line with the machine rotational b-axis and the z_{KS_1} is one direction with the machine rotational c-axis. KS_1 and KS_2 have the same point of origin and are mutually able to convert using the Euler angles α and γ. The remaining Euler angle β corresponds the angle q_2 as rotation about the b-axis. The rotation about the c-axis by angle q_1 corresponds the Euler angle γ so that the axis of direction in z is the z-axis and the axis in direction of y is defined as y-axis. For that the rotation order is important: It is mandatory to define q_2 at first, followed by q_1. It follows the equals:

$$q_1 = \gamma \tag{1}$$

$$q_2 = \beta \tag{2}$$

Let the coordinate system KS_3 the workpiece coordinate system. In order to determine the join angle q_3, q_4 and q_5 for the given coordinates x_{KS_3}, y_{KS_3} and z_{KS_3} (as target point given by the NC-program), these coordinates have to be converted from the workpiece coordinate system KS_3 to the machine coordinate system KS_0. At first there is the transformation to KS_2, than to KS_1 and finally to KS_0. The required equals are defined as:

$$x_{KS_2} = (\cos \alpha * x_{KS_3} + \sin \alpha * y_{KS_3}) + {}^3V_x^2 \tag{3}$$

$$y_{KS_2} = (-\sin \alpha * x_{KS_3} + \cos \alpha * y_{KS_3}) + {}^3V_y^2 \tag{4}$$

$$z_{KS_2} = z_{KS_3} + {}^3V_x^2 \tag{5}$$

α describes the angel of the x_{KS_2} and x_{KS_3}-axis rotating about the z-axis when their points of origins are overlapped as well as the linked orientation of the workpiece on the machine table. ${}^3V_x^2$ represents the displacement vector which has its direction from the point of origin of KS_3 to the point of origin of KS_2.

The next step is to transform the coordinates in the coordinate system KS_2 to KS_1. For that, it is notable to include the rotation of the coordinates when the machine table is turning, e.g. during the production caused by the NC-commands. For the transformation from the KS_2-system to the KS_1-coordinates, the Euler angles are used to build a transformation matrix. At first, the rotation about the z-axis occurs using angle γ followed by the rotation about the b-axis using angle β:

$$
{}^2Rot_1(\beta, \gamma) = \begin{pmatrix} \cos \gamma & -\sin \gamma & 0 \\ \sin \gamma & \cos \gamma & 0 \\ 0 & 0 & 1 \end{pmatrix} * \begin{pmatrix} \cos \beta & 0 & \sin \beta \\ 0 & 1 & 0 \\ -\sin \beta & 0 & \cos \beta \end{pmatrix} \tag{6}
$$

$$^2Rot_1(\beta, \gamma) = \begin{pmatrix} \cos\beta\cos\gamma & -\sin\gamma & \cos\gamma\sin\beta \\ \sin\gamma\cos\beta & \cos\gamma & \sin\gamma\sin\beta \\ -\sin\beta & 0 & \cos\beta \end{pmatrix} \qquad (7)$$

It follows for the coordination transformation for KS_1:

$$x_{KS_1} = (\cos\beta * \cos\gamma) * x_{KS_2} + (\sin\gamma * \cos\beta) * y_{KS_2} - \sin\beta * z_{KS_2} \qquad (8)$$

$$y_{KS_1} = -\sin\gamma * x_{KS_2} + \cos\gamma * y_{KS_2} \qquad (9)$$

$$z_{KS_1} = (\sin\beta * \cos\gamma) * x_{KS_2} + (\sin\beta * \sin\gamma) * y_{KS_2} + \cos\beta * z_{KS_2} \qquad (10)$$

With usage of the displacement vector $^1V_0$ the coordinates x_{KS_1}, y_{KS_1} and z_{KS_1} are able to transform in the coordination system KS_0:

$$q_3 = x_{KS_0} = x_{KS_1} + {}^1V_{0x} \qquad (11)$$

$$q_4 = y_{KS_0} = y_{KS_1} + {}^1V_{0y} \qquad (12)$$

$$q_5 = z_{KS_0} = z_{KS_1} + {}^1V_{0z} \qquad (13)$$

In order to identify the coordinates x_{KS_3}, y_{KS_3} and z_{KS_3} for the coordination system KS_3, the coordinates x_{KS_1}, y_{KS_1} and z_{KS_1} are identified at fist followed by the transformation to the system KS_2. For this, the transformation matrix (7) can be inverted.
It follows the equals:

$$x_{KS_1} = x_{KS_0} - {}^1V_{0x} \qquad (14)$$

$$y_{KS_1} = y_{KS_0} - {}^1V_{0x} \qquad (15)$$

$$z_{KS_1} = z_{KS_0} - {}^1V_{0x} \qquad (16)$$

$$^2Rot_1(\beta, \gamma) = \begin{pmatrix} \cos\beta\cos\gamma & \sin\gamma\cos\beta & -\sin\gamma \\ -\sin\gamma & \cos\gamma & 0 \\ \sin\beta\cos\gamma & \sin\beta\sin\gamma & \cos\beta \end{pmatrix} \qquad (17)$$

$$x_{KS_2} = (\cos\beta\cos\gamma) * x_{KS_1} - \sin\gamma * y_{KS_1} + (\sin\beta\cos\gamma) * z_{KS_1} \qquad (18)$$

$$y_{KS_2} = (\sin\gamma\cos\beta) * x_{KS_1} + \cos\gamma * y_{KS_1} + (\sin\beta\sin\gamma) * z_{KS_1} \qquad (19)$$

$$z_{KS_2} = -\sin\gamma * x_{KS_1} + \cos\beta * z_{KS_1} \qquad (20)$$

After the transformation of the coordinates from the coordination system KS_2 to KS_3, it follows:

$$x_{KS_3} = \cos(-\alpha) * \left(x_{KS_2} - {}^3V_{2x}\right) + \sin(-\alpha) * \left(y_{KS_2} - {}^3V_{2y}\right) \tag{21}$$

$$y_{KS_3} = -\sin(-\alpha) * \left(x_{KS_2} - {}^3V_{2x}\right) + \cos(-\alpha) * \left(y_{KS_2} - {}^3V_{2y}\right) \tag{22}$$

$$z_{KS_3} = z_{KS_2} - {}^3V_{2z} \tag{23}$$

The given model presents the theoretical concept to allow transformation between node-coordinates and given target coordinates from the NC-program under the restrictions of finite movements of the machine and the maximum of 5-machine-axis ordered in a kinematic chain that is connected in series.

3.4 Model for Use Case 2: Combination of Translation Vectors and Quaternions

The following model will show the determination of the inverse kinematic problem of the 5-axis-tooling machine using translation vectors and quaternions.

The transformation between the coordination systems KS_3 and KS_2 are performed by displacement vector ${}^3V_2$ and quaternion ${}^3Q_2$. For the definition of the quaternion ${}^3Q_2$, there is a change of sign of the angle α between KS_3 and KS_2:

$$^3Q_2 = \cos\frac{-\alpha}{2} + \sin\frac{-\alpha}{2} * k \tag{24}$$

$$P_{KS_2} = P_{KS_3} + 2 * {}^3Q_{2xyz} \otimes \left({}^3Q_{2xyz} \otimes P_{KS_3} + {}^3Q_{2w} * P_{KS_3}\right) + {}^3V_2 \tag{25}$$

The transformation from KS_2 to KS_1 will be performed by the total-quaternion. In order to use equals

$$w = \cos(\varphi/2) \tag{26}$$

$$(x, y, z)^T = \sin(\varphi/2) * \vec{D} \tag{27}$$

two rotation quaternions for the b- and c-axis are definable:

$$^2Q_{1.1} = \cos\frac{-\beta}{2} + \sin\frac{-\beta}{2} * k \tag{28}$$

$$^2Q_{1.2} = \cos\frac{-\gamma}{2} + \sin\frac{-\gamma}{2} * j \tag{29}$$

φ is a rotation angle in the range $\varphi \in [0, \pi]$, D describes the rotation axis with the unit vector $D = (D_x, D_y, D_z)^T$ (see [7, 10]), ${}^3Q_{2xyz}$ and ${}^3Q_{2w}$ describe transformation quaternions. The general nomenclature for quaternions is defined as Q. $P_{KS_{1,2,3}}$ describes a point including given coordinates in a defined coordination system KS.

The b-axis rotates about the y-unit vector and the c-axis rotates about the z-unit vector. For the angles, there are also a change of signs necessary.

Including the quaternion calculation rules for multiply quaternions, the total-quaternion is defined as:

$$^2Q_1 = \cos\frac{-\gamma}{2}\cos\frac{-\beta}{2} + \left(\cos\frac{-\gamma}{2}\sin\frac{-\beta}{2}\right)*j + \left(\sin\frac{-\gamma}{2}\cos\frac{-\beta}{2}\right)$$
$$*k + (\sin\frac{-\gamma}{2}\sin\frac{-\beta}{2})*i \qquad (30)$$

For the variables j, i, k is notable that there are elements from the complex numbers: j, i, k $\in \mathbb{C}$.

For the transformation of the coordinates of a point in the coordinate system KS_2 into KS_1, the following equal is necessary:

$$P_{KS_1} = P_{KS_2} + 2 * {^2Q}_{1xyz} \otimes ({^2Q}_{1xyz} \otimes P_{KS_2} + {^2Q}_{1w} * P_{KS_2}) \qquad (31)$$

For the transformation of the coordinates from KS_1 to KS_0, the equals (11), (12) and (13) in order to determine q_3, q_4, q_5 can be used. For the determination of q_1 and q_2 the eqals (1) and (2) are necessary.

4   Review Using Theoretical Scenarios

After comparison of the approaches from Sects. 3.3 and 3.4, the basic calculation steps are shown in Table 1. The result indicates that the less complex model using rotation matrix and translation vectors are sufficient for the low complex inverse kinematic such as the 5-axis machine model as an inverse kinematic chain. The basic calculation operation number of model 1 amounts $\sim 64\%$ compared to model 2. The model 1 fulfills the requirements for the research issue 1 more than model 2 for less complex kinematic problems. The results can be explained with the fact that the transformation between two coordination systems contain only one transformation matrix which is very operational in practice using displacement vectors and rotation matrix. For parallel inverse kinematic problems, the more complex approach using quaternions in combination with translation vectors would be more profitable to implement, because the number calculation steps using quaternions is less than operating with matrix-multiplication instead [6, 10]. In addition the quaternions are numerically more stable than Euler angles and there appears no singularities [10].

To ensure if the given models follow a logically machine behavior there will be conducted pilot-calculation containing given coordinate systems and target coordinates (points) which are initialized by the machine. The calculation should estimate, that the target coordinates are valid or how far the results determine invalid machine and production behavior. Figure 4 shows an example. The given target coordinates represent the result that P_1 is calculable and successful retractable for the machine mode. P_2 is determined as invalid result, because the coordinate is out of the accessible area of the machine (see Fig. 4).

Table 1. Overview of the basic computing steps to determine results

Approaches	Multiplication operations	Additions operations	Geometrical elementary function operations	Total result of basic operation number
Model use case 1	16	13	16	45
Model use case 2	32	28	10	70

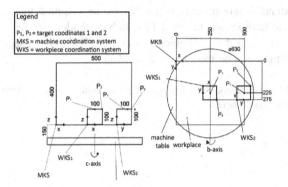

Fig. 4. Calculation scenario with two target coordinates (side and top view)

Under the role model of a real tooling machine[1] the two models offer high potential to determine invalid target positions of the tool paths and production operations, because the mathematical calculation contains a fraction of computing durations compared to the full simulation model, based on graphical presentation of the production process. The usage of inverse kinematic models under restrictions of the real machine is a useful chance to implement a successful pre-processing system in combination with the NC-parser. The pre-processing serves an estimation of near-valid setup positions of workpiece in the work area of the machine. That means, the search of valid positions without a visualization and waiting period until the simulation is finished can be spared and the probability that the results of the model represent valid positions could be increase.

Result for research issue 2: As input data, the x-,y- and z-coordinates are given by the NC-program and can be read in automatically. For rotation processes, it is more complicate to read in the specified NC-cycles. For that it is necessary to define rules for the automatically data processing. The defined cycle data from the NC-program could contain the circumstances that there have to recalculate in angle values for the Euler angles or related angle data.

---

[1] As role model it was used the 5-axis tooling machine "DMU 50" designed by Mori Seiki AG, Germany.

5 Conclusion and Outlook

This contribution introduces two inverse kinematic models to calculate the machine behavior under the restrictions of a real role model machine with the goal to identify valid positions for workpieces. The inverse kinematic models are kept as simple in order to prevent high calculation effort and it is shown that the usage of rotation matrix and translation vector is sufficient for the inverse kinematic as series chain. Quaternions and translation vectors is usable for more complex kinematic model e.g. parallel kinematic models. Research issue 1 will be fit for model 1 and for model 2 in order to address a high complex inverse kinematic problem. Research issue 2 leads to the result that there are extended data processing for NC-cycles required that contains rotation commands. For the future work, the model has to be extended (see Sect. 3.1) for further tests in combination with the PSO-algorithm and the NC-parser especially the evaluation by the tooling machine model (see Sect. 3.1).

Acknowledgments. We thank Silas Lummer for the model preparation and his support to execute the tests and experiments.

References

1. Reisch, R.-E., Weber, J., Laroque, C., Schröder, C.: Asynchronous optimization techniques for distributed computing applications. In: Tolk, A., Padilla, J.J., Jafar, S. (eds.) Spring Simulation Multi Conference 2015 Proceedings, 48th Annual Simulation Symposium, Alexandria, Virginia, vol. 47. No. 2, pp. 49–57. Institute of Electrical and Electronics Engineers, Inc. (2015)
2. Weber, J.: A technical approach of a simulation-based optimization platform for setup-preparation via virtual tooling by testing the optimization of zero point positions in CNC-applications. In: Yilmaz, L., Chan, W.K.V., Moon, I., Roeder, T.M.K., Macal, C., Rossetti, M.D. (eds.) Proceedings of 2015 Winter Simulation Conference, Huntington Beach, CA, USA (2015)
3. Weber, J., Mueß, A., Dangelmaier, W.: A simulation based optimization approach for setting-up CNC machines. In: Doerner, K.F., Ljubic, I., Pflug, G., Tragler, G. (eds.) Operations Research Proceedings 2015. ORP, pp. 445–451. Springer, Cham (2017). doi:10. 1007/978-3-319-42902-1_60
4. Mueß, A., Weber, J., Reisch, R.-R., Jurke, B.: Implementation and comparison of cluster-based PSO extensions in hybrid settings with efficient approximation. In: Niggemann, O., Beyerer, J. (eds.) Machine Learning for Cyber Physical Systems – Selected Papers from the international Conference ML4CPS 2015 – Technologien für die intelligente Automation, Lemgo, Germany, vol. 1, pp. 87–93. Springer, Heidelberg (2015)
5. Laroque, C., Weber, J., Reisch, R.-R., Schröder, C.: Ein Verfahren zur simulationsgestützten Optimierung von Einrichtungsparametern an Werkzeugmaschinen in Cloud-Umgebungen. In: Nissen, V., Stelzer, S., Straßburger, S., Firscher, D. (eds.) Proceedings of the Multikonferenz Wirtschaftsinformatik (MKWI) 2016, vol 3, Ilmenau, Germany, pp. 1761–1772. Monsenstein und Vannerdat OHG, Universitätsverlag Ilmenau (2016)
6. Weber, W.: Industrieroboter: Methoden der Steuerung und Regelung. Fachbuchverlag, Carl-Hanser-Verlag, Leipzig, Germany (2009)

7. Wenz, M.: Automatische Konfiguration der Bewegungssteuerung von Industrierobotern. Logos Verlag, Berlin, Germany (2008)
8. Siciliano, B., Khatib, O.: Handbook of Robotics. Springer, Heidelberg (2008)
9. Sieger, H.-J., Bocionek, S.: Robotik: Programmierung Intelligenter Roboter. Springer, Heidelberg (1996)
10. Husty, M., Karger, A., Sachs, H., Steinhilper, W.: Kinematik und Robotik. Springer, Heidelberg (1997)
11. Selig, J.M.: Geometric Fundamentale of Robotics, 2nd edn. Springer, Heidelberg (2005)

ISSB

Ransomware-Prevention Technique Using Key Backup

Kyungroul Lee[1], Insu Oh[2], and Kangbin Yim[2(✉)]

[1] R&BD Center for Security and Safety Industries (SSI), Soonchunhyang University,
Asan, South Korea
carpedm@sch.ac.kr
[2] Department of Information Security Engineering, Soonchunhyang University,
Asan, South Korea
{catalyst32,yim}@sch.ac.kr

Abstract. In this paper, a key-backup technique for the recovery of files that have been encrypted by ransomware is proposed. Ransomware interferes with the victim's system through the enactment of abnormal behavior, which is the locking of the victim's system or the encryption of the system or files. Ransomware writers require money from victims as a condition for the recovery of the encrypted files and systems that have been seized; accordingly, systems that are infected by ransomware cannot be repaired without a decryption key, making the employment of detection and recovery methods urgent. In this paper, a prevention technique for the backing up of encryption keys in a secure repository, and that can enable the recovery of ransomware-infected systems and ransomware-encrypted files. The proposed technique can be used to repair systems that have been infected by ransomware, thereby ensuring safety regarding such malicious codes.

Keywords: Ransomware · Key backup · Prevention · Big-data security

1 Introduction

As modern society is changing into an information society, a variety of the corresponding information are utilized and stored. In the past, users stored their personal information onto storage media such as hard disks, floppy disks, and CDs. While the computing environment and the network environment continually develop, remote-storage technologies such as Web hard drives and cloud services, whereby data as well as the computing environment such as software are used remotely, have emerged. With the development of the storage environment, the remote techniques are able to process mass data, and big data has appeared as a result. Big data is the derivation of the value of data through the collection, storage, management, and analysis of mass data.

Similar to other research fields, the most important element of big data is the data itself. In this respect, the analyzed results are valuable when data must be preserved without the need for an alteration; unfortunately, though, there is a problem regarding the insurance of the reliability of the data. Because many problems are caused by forgery data, a critical issue is the vulnerability of the system data that are stored, and a representative example is cyber-crime.

© ICST Institute for Computer Sciences, Social Informatics and Telecommunications Engineering 2017
J.J. Jung and P. Kim (Eds.): BDTA 2016, LNICST 194, pp. 105–114, 2017.
DOI: 10.1007/978-3-319-58967-1_12

Past cyber-crimes originated from curiosity, but it has developed into a tool for revenge, monetary purposes, and cyber warfare, and cyber-crime has become a kind of service; this is called "CaaS" (Crime as a Service) [1]. These services are being sold as products, and they are classified by consulting services as either botnet setups, infection and spreading services, botnet and rentals, and crimeware-upgrade modules. Concretely, a consulting service for a typical botnet setup is approximately 350 dollars to 400 dollars, while an infection and spreading service is approximately 100 dollars for every 1,000 that are installed. Big-data systems that comprise one of the various service-related data have therefore been exposed to threats, and the salience of this issue is further highlighted by the recent emergence of ransomware. In February 2016, one of the most notorious ransomware attacks occurred; due to this, Hollywood Presbyterian Medical Center in the U.S. did not access patient data because of the ransomware-encrypted files that had infected the medical organization's electronic medical-record system [2].

In this paper, a prevention technique for user PCs as well as a variety of systems such as big-data systems is therefore proposed to provide protection from ransomware-based cyber-crime. For the proposed technique, the encryption key is monitored and backed up in a secure repository. It has been confirmed that the use of the proposed technique provides safety from the threats that are caused by ransomware for data-storage systems.

2   Related Works

Related Works describes an overview of ransomware and previous countermeasures.

2.1   Overview of Ransomware

"Ransomware" is the name of the phenomenon here, and it is a new compound word comprised of "ransom" and "malware." The origin of the name is the act of demanding money from users whose information is being held "hostage" on a computer [5]. Ransomware is a kind of Trojan horse, as it penetrates systems so that it can seize or encrypts the files and resources on the system. Money is typically required by the cyber-criminal to release the held-hostage data [3, 4].

The ultimate purpose of ransomware is a monetary one. While users typically do not want to pay malicious attackers, they are left with no choice because their systems have been seized. The payment elements are classified by the user's education, the complexity level of the malware, and the urgency of the recovery, and users pay according to the correlation between these elements [4].

Regarding the domination of a system, data is one of the important factors of ransomware, while another important factor is the payment method. In terms of the payment methods, the following classifications apply: online monetary, irregular funding, and online shopping. Online monetary is a transmission method for which users transfer money to a designated online account using an online payment service such as PayPal, E-Gold, Bitcoin, or Webmoney [4, 6]. For irregular funding, users transfer illegal funds using legitimate businesses and services [4]. The online-shopping payment method involves an inducement by the attacker to the user for the purchase of goods from designated websites or web services. All of these payment methods are also utilized for money laundering because legitimate payments are used to provide camouflage for illicit activities [4].

Ransomware writers request either of the above-described payment methods by storing a text file such as a Readme.txt, or by displaying messages in pop-up windows. Examples of the payment-request methods are as follows [6]:

Ransomware first emerged in "PC CYBORG/AIDS information Trojan" in 1989, but this technique was not frequently used by malware writers, leading to its abandonment for almost 15 years. Nevertheless, through the widespread increase of Internet usage and e-commerce activities, ransomware was reborn in 2005 [4] (Table 1).

Table 1. Examples of payment-request methods

Ransomware name	Seizure type	Payment method
Trojan.Pluder.a	Hiding	Remit designated bank
Arhiveus	Encryption	Purchase pharmaceutical products from Russian websites
Trojan.Randsom.A	Deletion	Remit Western union
Trojan.Cryzip	Compression	Remit designated E-Gold account
Trojan.PGPCode	Encryption	Remit designated E-Gold account

The ransomware revival was mainly active in the U.S., but it gradually became widespread throughout the world. In the case of South Korea, ransomware emerged in April 2015. Notably, CryptoLocker was a new type of ransomware that was mostly active in the U.S. in 2013, and is now known as the most notorious and influential type of ransomware. According to ZDnet, ransomware writers have extracted approximately 27 million dollars in revenue from victims [8]. During the second quarter of 2013, 350,000 samples were found [9], and 14 kinds of ransomware appeared from January 2014 until September 2015 [7]. Regarding recent ransomware trends, 600,000 cases of ransomware were detected during the fourth quarter of 2015 [10].

The ransomware-infection process consists of five steps. The first step is the seeking out of the victim, for which ransomware writers propagate their ransomwares using spam mail or other propagation methods. The second step is the execution step, whereby the propagated ransomware is executed using social-engineering methods without the targeted user's knowledge; for example, CryptoLocker is disguised as a PDF icon when it is actually an executable file, so users execute ransomware by confusing the PDF file because the Windows operating system does not display file extensions by default. When ransomware is running, it generates a session key and an IV for communication with the writer. The third step is the generation of the key that encrypts the files of the system. In the case of public-key encryption, ransomware generates a secret key, so that the generated secret key is encrypted and sent to an attacker based on the received public key. Afterward, ransomware encrypts the files on the victim's system based on the generated secret key. In the case of secret-key encryption, ransomware generates a secret key for the encryption, so that the generated secret key is encrypted and sent to the attacker based on the generated session key. Afterward, the ransomware encrypts the files on the victim's system based on the generated secret key. The fourth step is the actual encryption, for which the ransomware actually encrypts the files based on the generated encryption key. The last step is the display of the message requesting payment that is either a stored text file or is displayed on the victim's screen [9].

According to Kaspersky, when it is operated in the above way, ransomware is exceedingly dangerous because the encrypted files cannot be recovered [6]. Numerous researchers have therefore engaged in studies to counteract ransomware threats, and the surveyed results of previous countermeasures are subsequently described.

2.2   Previous Countermeasures

Traditional anti-virus products generally detect and cure malicious codes based on signatures; for this reason, it is difficult to detect newly produced malicious codes. In particular, ransomware is very difficult to repair due to the post-infection encryption. In this section, the countermeasures that have been previously researched to solve these problems are described.

File-based detection method: This method detects the specific signatures of malicious actions in the particular format that is used by an operating system; for example, a PE (Portable Executable) file. The advantage here is a fast detection time; however, false negatives can sometimes occur, and it is difficult to detect new formats or unknown malicious codes [11].

System-based-behavior detection method: This method detects the malicious behaviors on a computer system, and integrity checking and behavior blocking are also carried out. Integrity checking is the checking of a system or files periodically to verify the integrity of the files based on the hash value of the execution files and the directories of a clean system. Behavior blocking is the monitoring of the entire behavior of a system; therefore, when malicious behaviors are detected, they are blocked by an anti-virus product according to the checked result after the tracing process [7].

Resource-based-behavior detection method: This method is the monitoring of specific resources for the detection of malicious behaviors. The targets of the monitored resources include CPU usage and I/O usage, among others. To detect the malicious behaviors, an anti-virus product collects the information of these resources on a clean system over a long duration. The product then determines malicious behaviors when exceptions are detected based on the collected information [7].

Connection-based-behavior detection method: This method is the monitoring of connection statuses. When connections are required, a user accepts or rejects connections. In the case of public-key-based ransomware, the ransomware receives a secret encryption key from the attacker's server such as the C&C server; accordingly, the installed ransomware tries to connect the server during this process. If an anti-virus product blocks the connections, ransomware does not encrypt the files or the system because it does not receive the key for the encryption. This method is therefore able to preserve the safety of a user's data and system [9].

Reverse engineering: This method is the recovery of encrypted files or an encrypted system based on the discovery of the key that is stored in the ransomware using reverse engineering. The advantage of this method is the repairing capability of the encrypted files when an anti-virus product does not detect the ransomware. Nevertheless, the drawback is the inability of the method to apply ransomware or an encrypted file that does not contain a decryption key [7].

3 Proposed Prevention Technique

As described in Sect. 2, an attacker requests a monetary sum while he/she is using a variety of ransomwares to hold a victim's system or files hostage; however, the prevention and recovery methods for the solving of this type of threat that have been introduced by security experts are limited in reality. One reason for this status is the need for a decryption key to recover a system or files, while another reason is the dependence on the payment of the ransom for the receipt of the decryption key that is stored in the attacker's system or server; for these reasons, victims suffer significant damages [6]. A countermeasure is therefore proposed in this paper, for which the key is backed up to recover the encrypted system and files.

3.1 Concept and Structure

Ransomware writers need the following requirements to run their ransomware successfully. First, a way to penetrate the victim's system successfully is needed. Users commonly install anti-virus programs to protect their systems; for this reason, writers penetrate victims' systems by using unknown vulnerabilities or known vulnerabilities such as the zero-day attack to avoid detection from anti-virus programs. Second, after penetration, ransomware should not be detected by an anti-virus program during the performance of malicious features such as the encryption and rootkit. After penetration, if ransomware does not implement the encryption, writers are unable to request money; therefore, writers want to stealthily execute features such as the encryption, and this technique is rootkit [12]. Third, writers use an implemented code that has been written by them or a library that is provided by the operating system to perform the encryption feature successfully. The encryption feature is the most important part of ransomware, because if it does not work or is not designed properly, a number of problems arise for the ransomware writers. One example here is the exposure of the encryption and decryption keys, while another example is the recovery capability when the ransomware does not implement the encryption properly. In the case of the implemented encryption feature, ransomware occasionally does not work on a specific system, so writers use cryptography libraries for reliability. Lastly, the number of keys that are generated for the encryption is equivalent to the number of the infected systems; in this case, writers generate one key equally for reliability because it is difficult for writers to manage the generated keys.

3.2 Assumption

In this paper, both the concept and the structure of the proposed countermeasure are described based on the above-defined requirements; here, an especial focus is placed on the third and fourth requirements. In the case of the encryption feature, it is assumed that ransomware writers use cryptography libraries due to a reliability that is greater than that from the implementation of the codes by the writers themselves. For the Windows operating system, a variety of commercial cryptography libraries are provided, and the newly adopted CNG library that was adopted during the development of Windows Vista, Windows 7, and Windows 8 is especially notable [13, 14]. Accordingly, it is assumed that ransomware uses the CNG library or downloads encryption codes from outside of the server, and then the codes encrypt the victim's files. In the case of the encryption key, when ransomware generates the key inside or receives it from outside of the server, the ransomware must generate or import the key. For this feature, ransomware uses specific functions of the CNG library; therefore, the proposed concept serves as the basis for the defining of the utilized functions with respect to the encryption on the CNG library.

Generation and import functions for the key: The BCryptGenerateSymmetricKey and BCryptGenerateKeyPair functions for the generating keys, and the BCryptImportKey and BCryptImportKeyPair functions for the importing keys.

Encryption and Decryption: The BCryptEncrypt function for the encryption and the BCryptDecrypt function for the decryption.

For this paper, it is assumed that ransomware encrypts the victim's files using the CNG library; therefore, the prevention program hooks both the key-generation and the key-import functions for the backing up of the keys, and then the obtained keys are stored in a safe repository of the victim's system, or they are sent to an authentication server or a certificate server. Nevertheless, if the steps for the key generation and import

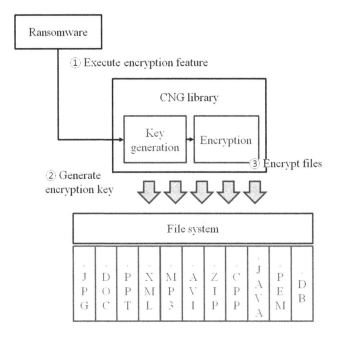

Fig. 1. The whole structure of the ransomware-encryption step for which the CNG library is used

are not executed by the ransomware, the encryption is based on the inside key that is in the ransomware file, and the prevention program backs the keys up during the encryption step. Figure 1 shows the entire structure of the ransomware-encryption step for which the CNG library is used.

Step 1. Ransomware penetrate the victim's system and executes the encryption feature to encrypt the files of the victim's system.

Step 2. Ransomware loads the CNG library to execute the encryption, followed by its generation of the encryption key. When the key is generated successfully, the generated key is sent to the attacker.

Step 3. Ransomware encrypts files such as .JPG and .DOC that are on the victim's system.

3.3 Prevention Technique

The structure of ransomware is described above, whereby it calls the BCryptGenerate-SymmetricKey function of the CNG library to generate the secret key, or it calls the BCryptGenerateKeyPair function of the CNG library to generate the public/private key-pair. The prevention program searches all of the programs that are loaded on the CNG library, and then hooks the BCryptGenerateSymmetricKey function to back up the secret key and the BCryptGenerateKeyPair function to back up the public/private key-pair; then, when the ransomware generates the key, the hooked codes are executed by the prevention program. The codes extract the encryption key and store it in the safe repository. If the ransomware does not generate the key due to its inclusion inside, it uses the key during the encryption step so that the prevention program stores it at that time. To import the key, the ransomware calls the BCryptImportKey function in the case of secret-key cryptography, or it calls the BCryptImportKeyPair function in the case of public-key cryptography; for this reason, the prevention program hooks these functions when they are called by the ransomware, and then the program stores them in the safe repository. Nevertheless, if the prevention program does not obtain the encryption key during the above steps, the program extracts the key during the encryption step. Concretely, to encrypt the files, ransomware calls the BCryptEncrypt function so that the prevention program hooks the function and stores it in the safe repository. Figure 2 shows the entire structure of the proposed prevention technique.

Step 1. Ransomware penetrates the victim's system and executes the encryption feature to encrypt the files of the victim's system; for this concept, the prevention program passes the encryption feature instead of blocking it. In terms of the prevention program, it does not determine whether the running program is a clean program such as the IE (Internet Explorer) and Outlook program or ransomware.

Step 2. When ransomware calls the key-generation and key-import functions of the CNG library, the hooking code passes the key onto the prevention program, and then the code passes the execution control onto the ransomware. The delivered key is stored in the safe repository by the prevention program that can be one module of an anti-virus program. The extracted keys are stored in the victim's safe repository or sent to the outside of an authentication server or a certification server. The received or stored keys are stored securely using password-based authentication or the certificate-based access-control technique to prevent exposure to unauthorized persons. Additionally, to improve security, the keys can be re-encrypted based on the shared keys, which are received from reliable agencies such as certification agencies and investigative agencies. When the files of the victim's system are encrypted by ransomware, the system is therefore able to recover through a decryption for which the extracted keys are used by commissioning investigation agencies or certificate agencies.

Step 3.   If the prevention program does not obtain the encryption key, the program extracts the key during the encryption step. The key-storage and key-recovery processes are equal to those of Step 2.

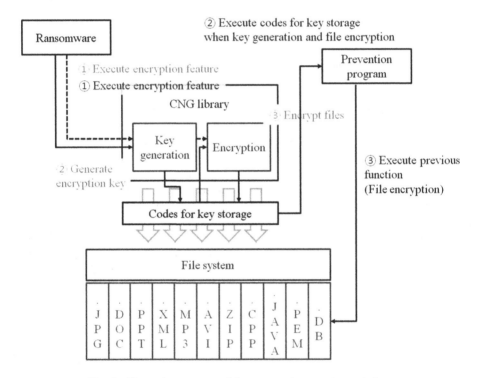

Fig. 2.   The entire structure of the proposed prevention technique

4 Conclusion

In this paper, the proposed ransomware-prevention technique uses a key-backup process to recover the encrypted files of a system that has been infected by ransomware. Ransomware uses abnormal behavior, which is the locking of the victim's system or the encryption of system or files, to interfere with a victim's system; therefore, a key-backup technique for which the encryption key is stored in a safe repository is proposed in this paper. For reliability, ransomware writers use the cryptography libraries that are provided by operating systems to execute the encryption feature; these codes call functions for the generation and importation of keys, and encryption of the files. When ransomware calls these functions, the prevention programs back the keys up in the safe repository, and the program subsequently recovers the encrypted files and the system using the stored keys. This study has confirmed that the proposed countermeasure can be used for the recovery of files and systems that have been infected by ransomware.

Acknowledgment. This research was supported by the Basic Science Research Program through the National Research Foundation of Korea (NRF) that is funded by the Ministry of Education (NRF-2015R1D1A1A01057300) and the MSIP (Ministry of Science, ICT and Future Planning), Korea, under the ITRC (Information Technology Research Center) support program (IITP-2016-R0992-16-1006) supervised by the IITP (Institute for Information & communications Technology Promotion).

References

1. Manky, D.: Cybercrime as a Service: a very modern business. J. Comput. Fraud Secur. **2013**(6), 9–13 (2013)
2. Everett, C.: Ransomware: to pay or not to pay? J. Comput. Fraud Secur. **2016**(4), 8–12 (2016)
3. Xin, L., Qinyu, L.: Ransomware: a new cyber hijacking threat to enterprises. In: Handbook of Research on Information Security and Assurance, IGI (2009)
4. Giri, B.N., Jyoti, N.: McAfee AVERT, The Emergence of Ransomware, AVAR (2006)
5. Gazet, A.: Comparative analysis of various ransomware virii. J. Comput. Virol. **6**(1), 77–90 (2008)
6. Liao, Q.: Ransomware: a growing threat to SMEs. In: Conference Southwest Decision Science Institutes (2008)
7. Song, S., Kim, B., Lee, S.: The effective ransomware prevention technique using process monitoring on android platform. J. Mobile Inform. Syst. **2016**, 9 (2016)
8. Violet, B.: CryotoLocker's crimeware: a trail of millions in laundered Bitcoin, ZDNet, December 2013
9. Ahmadian, M.M., Shahriari, H.R., Ghaffarian, S.M.: Connection-monitor & connection-breaker: a novel approach for prevention and detection of high survivable ransomwares. In: Conference Information Security and Cryptology, pp. 79–84, September 2015
10. McAfee, McAfee Labs Threats report. http://www.mcafee.com/us/resources/reports/rp-quarterly-threats-mar-2016.pdf
11. Kim, D., Kim, S.: Design of quantification model for ransom ware prevent. World J. Eng. Technol. **3**, 203–207 (2015)
12. Kim, S., Park, J., Lee, K., You, I., Yim, K.: A brief survey on rootkit techniques in malicious codes. J. Internet Serv. Inform. Secur. **2**(3/4), 134–147 (2012)

13. Lee, K., Lee, Y., Park, J., You, I., Yim, K.: Security issues on the CNG cryptography library (Cryptography API: Next Generation). In: Conference Innovative Mobile and Internet Services in Ubiquitous Computing, pp. 709–713, July 2013
14. Lee, K., You, I., Yim, K.: Vulnerability analysis on the CNG crypto library. In: Conference on Innovative Mobile and Internet Services in Ubiquitous Computing, pp. 221–224, July 2015

Multi-level Steganography Based on Fuzzy Vault Systems and Secret Sharing Techniques

Katarzyna Koptyra and Marek R. Ogiela[(⊠)]

Faculty of Electrical Engineering, Automatics,
Computer Science and Biomedical Engineering,
AGH University of Science and Technology,
30 Mickiewicza Ave., 30-059 Krakow, Poland
{kkoptyra,mogiela}@agh.edu.pl

Abstract. This paper presents a new technique of secret sharing in fuzzy vault system with use of multi-level steganography. Every hidden information on each level is linked with individual key used in embedding and revealing stages. Higher level secret is a share which is needed to reconstruct concealed data. It is possible that different shares contain various amount of information, what means that some of them may be more or less privileged. As a result, the secret can be decoded correctly in a number of ways. Therefore the main objective of this paper is to propose a system in which some users may hide additional shared secret in an inconspicuous manner.

Keywords: Secret sharing · Steganography · Fuzzy vault · Information hiding

1 Introduction

Nowadays some pieces of information are so important that they should be divided into parts and distributed to a few people as it is dangerous to accumulate so much secret data and responsibility in the hands of one person. This need resulted in continuous development of secret sharing algorithms. Some of these methods allow to graduate rights, what means that the secret can be correctly decoded in many ways, for example with use of lower number of more privileged shares or higher number of less privileged ones.

In practice there are cases where in multi-users system only a small group of people should have access to the secret message and remain users are not supposed to be aware of the existence of this information. The first solution which comes to mind is to create a brand new system available only for authorized users. Two main disadvantages of this idea are necessity of maintaining one more system and also problems with keeping its existence private. Thus one can try to create from scratch a new system, which first and overt functionality is to protect data owned by each individual user and second, covert feature is to keep hidden secret that is shared between authorized group of users.

This paper is the first attempt of initial proposal of such system. In its assumptions, the mechanism is built on three pillars: secret sharing algorithm, multi-secret fuzzy vault scheme, and multi-level steganography. These bases were described in Sect. 2,

© ICST Institute for Computer Sciences, Social Informatics and Telecommunications Engineering 2017
J.J. Jung and P. Kim (Eds.): BDTA 2016, LNICST 194, pp. 115–121, 2017.
DOI: 10.1007/978-3-319-58967-1_13

while the system itself is presented in Sect. 3. Finally some conclusions and summary can be found in Sect. 4.

2   The Fundamentals of the System

To understand how proposed system works, the first important thing is to become familiar with its theoretical basis, which are: secret sharing, multi-secret fuzzy vault and multi-level steganography.

2.1   Secret Sharing

Secret sharing is an idea that allows to divide a secret into several parts (called shares or shadows) and distribute them to authorised participants [1]. To reconstruct the secret, a certain number of users have to cooperate and join their pieces. The number of shares below the threshold is unfeasible to make reconstruction [2]. Usually the shared secret is a number [2–4], sometimes may take different forms, like matrices [5, 6] or images [7–10]. It is possible to make hierarchical [4] or multistage systems [11]. In general, various algorithms may have additional features, for example larger number of shared secrets, ease of adding a new user, ease of changing the secret, verifiability or shadow reusability.

This type of information management has two main aims. Firstly, it removes single point of failure flaw, because the secret can be reconstructed even if some participants lose their shares. Secondly, it prevents from concentration of too much potential in the hands of one user, because no one can recover the secret alone.

2.2   Multi-secret Fuzzy Vault

Fuzzy vault is a type of cryptosystem based on polynomial reconstruction [12]. The whole system is made of a great number of points, from which some are genuine and remaining are noise. To unlock hidden information, right points have to be identified with use of a key that is in form of unordered set of numbers. On the basis of chosen points the user can reconstruct a polynomial with encoded secret (a number). Therefore the main idea of this system is to protect an information by placing it in a very noisy environment. In vault creation process some coefficients of polynomial are dependent on secret value, others are selected randomly. This polynomial is then used to obtain genuine points by evaluate the formula on all key values. Further lots of chaff points (not lying on polynomial) are generated randomly or with use of more sophisticated algorithms [13, 14].

The interesting feature of fuzzy vault scheme is its error tolerance. The user can decode the secret with use of similar, but not identical key. However, both keys have to overlap substantially, otherwise the obtained genuine points are not sufficient to reconstruct the polynomial. This property in combination with order-invariance opens the possibility of using keys derived from biometric traits [15].

Multi-secret fuzzy vault is an extension of described scheme which allows to hide more secrets [16]. Every number is encoded in its own polynomial and has individual key. The important assumption is that all keys are disjunctive (as they are still unordered sets). Another property that distinguishes multi-secret version from original one is false points generation method. In extended scheme chaff points cannot be positioned on any of polynomials. All remaining characteristics of fuzzy vault are preserved, including error tolerance and order invariance. It is worth to mention that above rules are suitable for single user with many secrets or for multi-users system, in which each participant has his own key.

2.3   Multi-level Steganography

Steganography is a technique of information hiding in which some secret data is embedded in an inconspicuous medium called container (carrier, cover) [17]. The main assumption is to prevent disclosure of the secret, because when its existence is revealed, the whole system failed [18]. Therefore every steganographic method is supposed to be highly undetectable against unauthorized recipients. Currently there is a wide selection of algorithms and various covers in this interesting field [19].

Multi-level steganography is a branch of such science which provides different levels of hiding. In other words, a container with embedded data become new message and is hidden in an another carrier. Usually each subsequent level requires more capacity. To extract the last information, one should first decode every secret from all previous levels. More about multi-level steganography may be found in [20, 21].

3   Multi-level Fuzzy Vault System for Secret Sharing and Steganography

This section describes main elements of the newly proposed system. At the beginning it is worth to remain main assumptions, which are as follows. The system is destined for multiple users and has two levels of information hiding. First level is available for every user who wants to protect one numeric secret. Keys are in form of disjunctive unordered sets. Second level is known only to an authorized group of users and stores one shared secret. To decode this information, a sufficient number of users has to cooperate and combine their parts. Two things are required for computing a share: 1st level secret reconstruction and 2nd level key (numeric).

Below are explained hiding and reconstruction phases with details of overt and covert levels.

3.1   Hiding Phase

Every normal user applies Algorithm 1 for locking his own secret. The covert level is known only to specific group and requires special method described in Algorithm 2. This technique conceals two secrets as they are related. In depicted algorithms the assumption is that all polynomials are of degree n. All sets are indexed from 1, for instance 3-element set S can be denoted as (S_1, S_2, S_3).

Algorithm 1. Hiding secret (normal user)

Input: s – secret (numeric), k – key (set of numbers)
Output: v – set of points
1. Compute l as length of k
2. Randomly choose coefficients $a_n, a_{n-1}, ..., a_1$
3. Create a polynomial $p(x) = a_n x^n + a_{n-1} x^{n-1} + ... + a_1 x + s$
4. Evaluate p on elements of k, in other words compute $p(k_1), p(k_2), ..., p(k_l)$
5. Add all computed points $p(k_1), p(k_2), ..., p(k_l)$ to v
6. Return v

For authorized users, the polynomial p is generated in a different way. It is based on temporary polynomial q which conceals higher level secret. The difference between these two formulas become second level key. Again p and q have the same degree n which is equal the length of second level secret.

Algorithm 2. Hiding secret (initiated user)

Input: s – 1st level secret (numeric), k – 1st level key (set of numbers), S – share (set of numbers)
Output: v – set of points, K – 2nd level key
1. Randomly choose coefficient a
2. Create polynomial $q(x) = a(x - S_1)(x - S_2)...(x - S_n)$
3. Compute canonical form of polynomial $q(x) = ax^n + ab_{n-1}x^{n-1} + ... + ab_1x + ab_0$
   where:
   $b_{n-1} = S_1 + S_2 + ... + S_n$
   $b_{n-2} = S_1 S_2 + S_1 S_3 + ... + S_1 S_n + S_2 S_3 + ... + S_{n-1} S_n$
   $...$
   $b_1 = S_1 S_2 ... S_{n-1} + S_1 S_2 ... S_{n-2} S_n + ... + S_2 S_3 ... S_n$
   $b_0 = S_1 S_2 ... S_n$
   Note: $b_{n-1}, b_{n-3}, ...$ are negative. So for odd n b_0 is negative and for even n b_1 is negative.
4. Compute $K = ab_0 - s$ (2nd level key)
5. Compute l as length of k
6. Create polynomial $p(x) = ax^n + ab_{n-1}x^{n-1} + ... + ab_1x + s$
7. Evaluate p on elements of k, in other words compute $p(k_1), p(k_2), ..., p(k_l)$
8. Add all computed points $p(k_1), p(k_2), ..., p(k_l)$ to v
9. return K, v

After computing sets of points for all participants and keys for authorized users, chaff points are generated. There is one limitation in comparison to original fuzzy vault scheme. Here false points cannot lie on any of polynomials p. It does not matter if they are located on a temporary polynomial q from algorithm 2. Finally all genuine and chaff points are gathered together to form a vault.

3.2   Reconstruction Phase

First level secret reconstruction is identical as in fuzzy vault scheme [12]. In short, the participant uses his key k to choose genuine points from entire vault. Then these points are used to reconstruct polynomial p. The secret s is read from free term of p. Obtaining data from second level is more complicated, therefore it is presented in algorithm 3.

Algorithm 3. Share reconstruction

Input: V – set of all points, k – key, K – 2nd level key
Output: S – share (set of numbers)
1.   Use k to filter genuine points from V
2.   Reconstruct polynomial p
3.   Compute polynomial $q(x) = p(x) + K$
4.   Write polynomial q in factored form $q(x) = a(x - S_1)(x - S_2)...(x - S_n)$
5.   Return S

As can be seen, each share is in form of set of numbers. This gives a chance of using them as a key in another fuzzy vault in which the secret is recovered as described in [12]. This selection is justified by an opportunity of reusing a part of the system as the method is identical as 1st level secret reconstruction (overt functionality of the system).

At the end of the section, it should be explained how it is possible to assign different privileges to users. The previous assumption was that all polynomials were of level n, which was also the length of each share. However, some parts may be weakened by placing redundant information inside. In other words, the reconstructed polynomial q has multiple roots and, as a result, some data in the share overlap. In a simple case with three users, key length = 3 and n = 2, we can assign following shares: User1 has (S1, S2), User2 has (S3, S3) and User3 has (S4, S4). In this scenario User1 has to cooperate with User2 or User3 (or with both of them), but User2 and User3 are not able to reconstruct the secret without User1, because their key has length 2.

4   Conclusions

This paper introduces a secret sharing system based on multi-secret fuzzy vault and multi-level steganography. Presented system offers two functionalities. First, overt, allows to protect one secret per user and is available for every participant. The second one is hidden and is known only to specific group of users who can share another concealed secret. Privileges can be graduated by creating shadows containing different amount of information. Additionally it is possible to set up various hierarchical structures and subgroups which have to cooperate, because some parts of the secret may be present only in one of them. Moreover one system may contain independent authorized groups that share different secrets and are not informed about remaining ones.

Acknowledgments. This work was supported by the AGH University of Science and Technology research Grant No 15.11.120.868.

References

1. Martin, R.: Introduction to secret sharing schemes. Computer Science Department, Rochester Institute of Technology, Rochester (2012)
2. Shamir, A.: How to share a secret. Commun. ACM **22**, 612–613 (1979)
3. Blakley, G.F.: Safeguarding cryptographic keys. In: National Computer Conference, pp. 313–317 (1979)
4. Zhenjun, Y.: Construction and application of multi-degree secrecy system based on threshold secret sharing. In: International Conference on Power System Technology, PowerCon 2006, pp. 1–4 (2006)
5. Bai, L.: A strong ramp secret sharing scheme using matrix projection. In: International Symposium on World of Wireless, Mobile and Multimedia Networks, WoWMoM 2006, pp. 652–656 (2006)
6. Wang, K., Zou, X., Sui, Y.: A multiple secret sharing scheme based on matrix projection. In: 33rd Annual IEEE International Computer Software and Applications Conference, COMPSAC 2009, pp. 400–405 (2009)
7. Chien, M.C., Hwang, J.: Secret image sharing using (t,n) threshold scheme with lossless recovery. In: 2012 5th International Congress on Image and Signal Processing (CISP), pp. 1325–1329 (2012)
8. Chang, C.C., Kieu, T.D.: Secret sharing and information hiding by shadow images. In: Third International Conference on Intelligent Information Hiding and Multimedia Signal Processing, IIHMSP 2007, vol. 2, pp. 457–460 (2007)
9. Bai, L.: A reliable (k, n) image secret sharing scheme. In: 2nd IEEE International Symposium on Dependable, Autonomic and Secure Computing, pp. 31–36 (2006)
10. Nag, A., Singh, J., Sarkar, D., Sarkar, P., Biswas, S.: Distortion free secret image sharing based on x-or operation. In: 2012 International Conference on Communications, Devices and Intelligent Systems (CODIS), pp. 286–289 (2012)
11. He, J., Dawson, E.: Multistage secret sharing based on one-way function. Electron. Lett. **30** (19), 1591–1592 (1994)
12. Juels, A., Sudan, M.: A fuzzy vault scheme. Des. Codes Cryptography **38**(2), 237–257 (2006)
13. Hani, M.K., Marsono, M.N., Bakhteri, R.: Biometric encryption based on a fuzzy vault scheme with a fast chaff generation algorithm. Future Generation Comp. Syst. **29**, 800–810 (2013)
14. Nguyen, T.H., Wang, Y., Nguyen, T.N., Li, R.: A fingerprint fuzzy vault scheme using a fast chaff point generation algorithm. In: 2013 IEEE International Conference on Signal Processing, Communication and Computing (ICSPCC), pp. 1–6. IEEE (2013)
15. Nandakumar, A.K.J.K., Pankanti, S.: Fingerprint-based fuzzy vault: Implementation and performance. IEEE Trans. Inf. Forensics Secur. **2**, 744–757 (2007)

16. Koptyra, K., Ogiela, M.R.: Fuzzy vault schemes in multi-secret digital steganography. In: 10th International Conference on Broadband and Wireless Computing, Communication and Applications, BWCCA 2015, Krakow, Poland, 4–6 November 2015, pp. 183–186 (2015)
17. Bailey, K., Curran, K.: Steganography – The Art of Hiding Information. BookSurge Publishing, New York (2005)
18. Cox, I., Miller, M., Bloom, J., Fridrich, J., Kalker, T.: Digital Watermarking and Steganography. Morgan Kaufmann Publishers, San Francisco (2008)
19. Subhedar, M.S., Mankar, V.H.: Current status and key issues in image steganography: A survey. Comput. Sci. Rev. **13–14**, 95–113 (2014)
20. Ogiela, M.R., Koptyra, K.: False and multi-secret steganography in digital images. Soft. Comput. **19**(11), 3331–3339 (2015)
21. Tang, M., Hu, J., Song, W.: A high capacity image steganography using multi-layer embedding. Optik – Int. J. Light Electron Opt. **125**(15), 3972–3976 (2014)

On Exploiting Static and Dynamic Features in Malware Classification

Jiwon Hong, Sanghyun Park, and Sang-Wook Kim[✉]

Hanyang University, Seoul, Republic of Korea
{nowiz,singhyun,wook}@hanyang.ac.kr

Abstract. The number of malwares is exponentially growing these days. Malwares have similar signatures if they are developed by the same group of attackers or with similar purposes. This characteristic helps identify malwares from ordinary programs. In this paper, we address a new type of classification that identifies the group of attackers who are likely to develop a given malware. We identify various features obtained through static and dynamic analyses on malwares and exploit them in classification. We evaluate our approach through a series of experiments with a real-world dataset labeled by a group of domain experts. The results show our approach is effective and provides reasonable accuracy in malware classification.

Keywords: Malware classification · Static analysis · Dynamic analysis · Feature extraction

1 Introduction

Malwares are continuously causing social and economic damages. Despite the combined effort of various companies in different countries such as Microsoft, Kaspersky, Ahnlab, and Avast, the number of malwares is growing more than ever in recent years.

There are two primary reasons for the fast growth of malwares: there are a large number of attackers who are developing new malwares continuously; due to a variety of tools, it is relatively easy to develop a malware. For years, a number of attackers are releasing new malwares while successfully evading laws. They often share their source codes, which could also be accidentally leaked out to the public. This enables other novice attackers to build new malwares without difficulty. A number of malwares with a similar method of attacking could come out in a short period of time. Moreover, new malwares may take advantage of polymorphism or metamorphism for evading detection [1–3].

Despite such efforts of attackers, it is highly likely that similar signatures can be found in two different malwares written by the same attacker when analyzed from diverse angles. It is because those malwares essentially share the same code or attack patterns. Using the shared signatures is a widely used technique for distinguishing malwares from ordinary programs [1,3].

© ICST Institute for Computer Sciences, Social Informatics and Telecommunications Engineering 2017
J.J. Jung and P. Kim (Eds.): BDTA 2016, LNICST 194, pp. 122–129, 2017.
DOI: 10.1007/978-3-319-58967-1_14

In general, the goal of malware detection is simply to identify malwares from benign programs. However, if we are able to define classes of malwares and to automatically classify a given sample to its class, it would be more helpful than mere detection. Class information could be utilized for deeper researches on malwares such as finding countermeasures, developing treatments, and digital forensics [2].

In this paper, we present a classifier which successfully determines the class of a given sample with a set of classes defined by domain experts. We exploit various static and dynamic features at the same time to achieve high classification accuracy. Also, we show the effectiveness of our classifier via experiments.

In Sect. 2, we describe the class definition and the dataset used in our research. Section 3 explains the various features of malwares extracted for classification. Section 4 describes the classification method. We show the effectiveness of our classifier via experiments in Sect. 5. We conclude our paper in Sect. 6.

2   Class Definitions and Dataset

The definition of classes could differ for the objectives of applications. In this research, we are focusing on finding the groups of attackers who release malwares. We used a class set defined by a group of experts based on the similar attackers. A classifier learned from such a class set should be useful for inferring which attacker group released a given sample.

We collected 1,086 malware samples in 2015. Each sample is labeled with one of 7 classes by domain experts. Each class has 155 instances on average. While the largest one has 434 instances, the smallest one has 24. Table 1 shows the number of instances in each class.

Table 1. Number of instances in each class

A	B	C	D	E	F	G
434	24	261	113	48	147	61

3   Feature Extraction

For classifying malwares, features that describe each malware should be extracted. Some of these features may be signatures important to identify each malware. Classification would be successful only if the most informative and distinguishing features are identified and extracted.

The core files of a malware usually are in the executable binary form. Both static and dynamic analyses could be used for analyzing such binary files. Static analysis is conducted directly on the binary file [1]. For example, extracting public methods or printable strings contained in the binary file and disassembling the binary file are part of the static analysis. Dynamic analysis is conducted by running the binary code inside a controlled sandbox such as virtual machines [1]. While running, we can investigate the detailed behavior of the malware, including network and filesystem usages.

As mentioned earlier, a large portion of malwares are polymorphic and/or metamorphic. In such cases, the classification would be accurate only if the features from the sufficiently various analyses are used. In this paper, we identified and extracted the following features for our classification.

3.1 Static Features

Static features are extracted by analyzing the binary file of a malware. However, static analysis can be seriously affected by the polymorphic characteristics of the malware. Static analysis should be conducted with the dynamic one at the same time.

Functions. We identify the groups of opcodes from a function chunk obtained by disassembling the given binary file as features. If there are reused code chunks in a new sample, we can find them by using these features. The function chunks have variable length and they also consume large storage capacity to be indexed. Therefore, we convert each chunk to a short fixed-length value using a one-way cryptographic hash function [6, 7].

Strings. We identify each printable strings contained in the binary code of the malware as a feature. Most of such strings could be meaningless; however, some of them can be used as useful signatures when attackers inserted the same string habitually as a symbol in their own malwares [4].

Imports and Exports. Malwares could be composed of multiple files. In the cases, the malware has the interfaces for calling shared libraries or providing its public methods. We identify such information including the names of the methods as imports and exports features. These features are less affected by the polymorphic or metamorphic characteristics of the malware, thereby being unique.

3.2 Dynamic Features

Dynamic analysis is more suitable for extracting features from polymorphic malwares than static analysis. In this research, we executed the given malware sample in a virtual machine and identified the following features based on the running behavior of the sample in its execution.

System API Calls. A large portion inside the binary code of the malware consists of the system API calls provided by the operating system. We identify such system API calls made by each malware as features [5–7].

Mutexes. Mutexes are used for locking of a memory location or preventing the execution of multiple instances of the malware. The names of mutexes might be the same normally if they are created by the same group of attackers. It is more likely if the codes are reused. We identify the names of mutexes used in each malware sample as features.

Networks. We analyze the network requests from the sample and identify their information as features. Typical examples of such information are DNS URLs and IP addresses. Malwares are likely to use the same DNS URLS and IP addresses if they are developed by the same attacker or developed for the same purpose.

Files. Most of malwares show the behavior of reading or writing some files on the filesystem. The behavior might be an act of attacking or concealing itself among normal files. In many cases, the address on the target filesystem indicates a location where a system file resides. The malwares developed by the same group of attackers may share these addresses because they are determined by the attackers heuristically. We identify such addresses as features as well.

Drops. Some malwares could extract a hidden file inside the executable binary code to store it in a filesystem or a memory location. They are also capable of downloading other files from the Internet. Such behaviors are called *drop*. We identify the storage locations, remote addresses, and hash values of such files as our features.

Keys. Microsoft Windows has a data storage called *registry* for storing settings for most of the Windows functions and other applications. A huge number of malwares designed for Microsoft Windows exploit it for attacking the system. We identify all the registry keys that a sample accesses as features.

4   Classification

In this paper, we employ Decision Tree and SVM [8], both well-known classification methods to build our classifier. Decision Tree builds a tree-shaped prediction model from the dataset by evaluating features that each item has. This prediction model is used to determine the class of a new sample.

Figure 1 shows an example of a decision tree prediction model, where each node represents a *decision*. Given a new sample, Decision Tree traverses to a child node which satisfies the condition for the sample, starting from the root node. The traverse is repeated until it reaches a leaf node. If Decision Tree reaches a leaf node, it classifies the given sample as the class label of the leaf node.

To learn such tree-shaped prediction model, Decision Tree first creates a new root node with a condition that splits the whole samples at best with a given splitting criterion. Second, it creates child nodes for the split set of samples and gives each of the nodes a condition that splits the samples best recursively. Finally, if each node cannot be split, then it makes the node a leaf node and labels the node with a class label that the most samples in the node have.

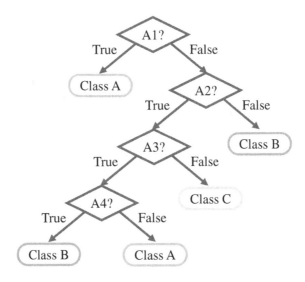

Fig. 1. An example of a decision tree.

We use *normalized information gain* [9] as the splitting criterion which is defined as follows:

$$Gain(D, A) = H(D) - \sum_{v \in V_A} \left(\frac{|D_v|}{|D|} \cdot H(D_v) \right)$$

$$SplitInfo(D, A) = - \sum_{v \in V_A} \frac{|D_v|}{|D|} \cdot log_2 \left(\frac{|D_v|}{|D|} \right)$$

$$GainRatio(D, A) = \frac{Gain(D, A)}{SplitInfo(D, A)}$$

Here, A is a feature; D is the set of all training samples for the node; V_A is the set of all possible values of A; D_v is the set of samples with the value v; and $H(D)$ represents the entropy of the set D. For each node in Decision Tree, the feature with the highest *GainRatio* is selected as the splitting feature.

Support Vector Machine (SVM) [8] is another well-known classification model. SVM generally shows high classification accuracy and has ability to construct complex models. Given a dataset, it tries to identify an optimal hyperplane or an optimal set of hyperplanes which discriminate two classes of data clearly. An optimal hyperplane is the one that has the largest margin to both classes of data and thus helps lower the errors of the classifier. An optimal hyperplane should satisfy the following condition:

$$W \cdot X + b = 0$$

W is a normal vector to the hyperplane with n elements where n is the number of attributes, X is a set of training tuples, and b represents the bias. To find an optimal hyperplane, we regard the equation above as a constrained convex quadratic optimization problem to find optimal W.

5   Evaluation

5.1   Experimental Setup

In our evaluation, we verify the accuracy of our classifier by training a Decision Tree classifier with our dataset. We used the WEKA library [10] to accomplish our task. We conducted a five-fold cross-validation for the evaluation. We used the following measures to evaluate the accuracy of our model:

$$Precision(C) = \frac{|L_C \cap R_C|}{R_C}$$

$$Recall(C) = \frac{|L_C \cap R_C|}{L_C}$$

$$F1(C) = 2 \cdot \frac{Precision(C) \cdot Recall(C)}{Precision(C) + Recall(C)}$$

Here, C is a class label; L_C is the set of test samples that have the class label C; and R_C is the set of test samples that are classified as C by our classifier. We used an weighted average over all the classes as overall accuracy, where the weight in each class indicates the number of samples in the class.

5.2   Results and Analyses

Table 2 is a confusion matrix M that shows the number of samples in each class has been classified to each class by the Decision Tree classifier. The rows show the actual classes of the samples and the columns show the predicted ones. For example, $M(A, A) = 388$ indicates our classifier correctly classifies 388 samples in Class A as Class A; $M(A, B) = 16$ indicates it misclassifies 16 samples in Class A as Class B. We can see that most of errors occurred in Class A that has the largest number of samples. Also, we observe that Class D shows poor accuracy because it does not have a sufficient number of samples. Notably, all the samples in Class F are correctly classified. We conjecture that Class F is the most distinguished one. Table 3 is a confusion matrix M for the SVM classifier. Table 3 also shows similar however slightly more accurate results.

Overall, the Decision Tree classifier shows the precision of 83.60%, the recall of 84.20% and the F1-measure of 83.40%; the SVM classifier shows the precision of 88.30%, the recall of 88.40% and the F1-measure of 88.20%. The results verify that our classifier performs effectively.

Table 2. A confusion matrix M (DT)

	A	B	C	D	E	F	G
A	388	7	2	5	22	0	9
B	16	41	0	0	4	0	0
C	4	0	14	1	5	0	0
D	21	1	0	11	13	0	2
E	27	3	0	1	225	0	4
F	0	0	0	0	0	113	0
G	19	2	0	1	3	0	122

Table 3. A confusion matrix M (SVM)

	A	B	C	D	E	F	G
A	406	4	0	6	8	0	9
B	12	45	0	0	2	0	2
C	0	0	16	3	4	0	1
D	11	1	0	24	9	0	3
E	23	0	2	2	229	0	4
F	1	0	0	0	0	111	1
G	15	1	0	2	0	0	129

6   Conclusions

In this research, we have addressed a malware classifier that exploits various features extracted from static and dynamic analyses. Our goal is to classify an author group of the given sample. We also have verified that our classifier provides reasonable accuracy via experiments with a real-life dataset.

As a further study, we plan to develop more accurate classifiers for malwares. Using more features such as a control flow [3] and call frequency or identifying a new set of features could be a good starting point towards this direction [6,7]. Another direction is to improve the training speed of our classifier by selectively use only the features that help classification. Finally, new class definitions for other applications can be considered as well.

Acknowledgement. This work was supported by (1), (2) the National Research Foundation of Korea (NRF) Grant funded by the Korean Government (MSIP) (2014R1A2A1A10054151 and 2015R1A5A7037751) and (3) the MSIP (Ministry of Science, ICT and Future Planning), Korea, under the ITRC (Information Technology Research Center) support program (IITP-2016-H8501-16-1013) supervised by the IITP (Institute for Information & communication Technology Promotion).

References

1. Islam, R., Tian, R., Batten, L.M., Versteeg, S.: Classification of malware based on integrated static and dynamic features. J. Netw. Comput. Appl. **36**(2), 646–656 (2013)
2. Tian, R., Batten, L.M., Islam, R., Versteeg, S.: An automated classification system based on the strings of trojan and virus families. In: Proceedings of the 4th International Conference on Malicious and Unwanted Software (MALWARE), pp. 23–30 (2009)
3. Cesare, S., Xiang, Y.: Classification of malware using structured control flow. In: Proceedings of the 8th Australasian Symposium on Parallel and Distributed Computing (AusPDC), pp. 61–70 (2010)
4. Islam, R., Tian, R., Batten, L.M., Versteeg, S.: Classification of malware based on string and function feature selection. In: Cybercrime and Trustworthy Computing Workshop (CTC), 2010 Second, pp. 9–17 (2010)
5. Park, Y., Reeves, D., Mulukutla, V., Sundaravel, B.: Fast malware classification by automated behavioral graph matching. In: Proceedings of the 6th Annual Workshop on Cyber Security and Information Intelligence Research (CSIIRW), pp. 45–49 (2010)
6. Chae, D., Ha, J., Kim, S.-W., Kang, B., Im, E., Park, S.: Credible, resilient, and scalable detection of software plagiarism using authority histograms. Knowl. Based Syst. **95**(1), 114–124 (2016)
7. Chae, D., Kim, S.-W., Cho, S.-J., Kim, Y.: Effective and efficient detection of software theft via dynamic API authority vectors. J. Syst. Softw. **110**, 1–9 (2015)
8. Han, J., Kamber, M.: Data Mining: Concepts and Techniques. Morgan Kaufmann, San Francisco (2006)
9. Quinlan, J.R.: C4.5: Programs for Machine Learning. Morgan Kaufmann Publishers, San Francisco (1993)
10. Hall, M., Frank, E., Holmes, G., Pfahringer, B., Reutemann, P., Witten, I.: The WEKA data mining software: an update. ACM SIGKDD Explor. Newsl. **11**(1), 10–18 (2009)

Distributed Compressive Sensing
for Correlated Information Sources

Jeonghun Park[1], Seunggye Hwang[2], Janghoon Yang[3], Kitae Bae[3],
Hoon Ko[4], and Dong Ku Kim[5(✉)]

[1] Department of Electrical and Computer Engineeing,
University of Texas at Austin, Austin, TX, USA
the20thboys@gmail.com
[2] LG Electronics, Anyang, Korea
pisces_sg@yonsei.ac.kr
[3] Department of New Media Contents,
Seoul Media Institute of Technology, Seoul, Korea
{jhyag,ktbae}@smit.ac.kr
[4] College of Information and Communication Engineering,
Sungkyunkwan University, Suwon, South Korea
skoh21@skku.edu
[5] School of Electrical and Electronics Engineering,
Yonsei University, Seoul, Korea
dkkim@yonsei.ac.kr

Abstract. The abstract should summarize the contents of the paper and should Distributed Compressive Sensing (DCS) improves the signal recovery performance of multi signal ensembles by exploiting both intra- and inter-signal correlation and sparsity structure. In this paper, we propose a novel algorithm, which improves detection performance even without a priori-knowledge on the correlation structure for arbitrarily correlated sparse signal. Numerical results verify that the propose algorithm reduces the required number of measurements for correlated sparse signal detection compared to the existing DCS algorithm.

Keywords: Compressive sensing · Distributed source coding · Sparsity · Random projection · Sensor networks

1 Introduction

Baron et al. [1] introduced Distributed Compressive Sensing (DCS), which exploits not just intra-, but also inter- correlation of signals to improve detection performance. In [1], they assumed Wireless Sensor Network (WSN) consisting of arbitrary number of sensors and one sink node, where each sensor carries out compression without cooperation of the other sensors and transmits the compressed signal to the sink node. At the sink node, it jointly reconstructs the original signals from the received signals. Here, a key of DCS is a concept of joint sparsity, defined as the sparsity of the entire signal ensemble. Three models have been considered as a joint sparse signal model in [1]. In the first model, each signal is individually sparse, and also there are common

© ICST Institute for Computer Sciences, Social Informatics and Telecommunications Engineering 2017
J.J. Jung and P. Kim (Eds.): BDTA 2016, LNICST 194, pp. 130–137, 2017.
DOI: 10.1007/978-3-319-58967-1_15

components shared by every signal vector, called common information. In the second model, all signals share supports, which means the locations of the nonzero coefficient. In the third model, although no signal is sparse itself, they share the large amount of common information, which makes it possible to compress and recover the signals using CS. The third model can be considered as a modified version of the first model. The second joint sparsity model (JSM-2) has been actively explored in many existing literatures [2–7]. A joint orthogonal matching pursuit (JOMP) for DCS was proposed to improve the target detection performance of MIMO radar [4]. A precognition matching pursuit (PMP) which used the knowledge of common support from Fr´echet mean was proposed to reduce the number of required measurements in WSNs [5]. DCS was shown to be feasible for a realistic wireless sensor WSNs by implementing on real commercial off-the-shelf (COTS) hardware with providing good trade-off between performance and energy consumption [6]. Exploiting common information across the multiple EGS signals, simultaneous orthogonal matching pursuit (SOMP) for DCS with learned dictionary was shown to provide accurate reconstruction with the reduced number of measurements [7]. However, to the best of authors' knowledge, the first model (JSM-1) has been studied relatively little. In addition, a limited ensemble of signals that have single common information is considered in most cases.

In this paper, we propose a generalized (GDCS). While the key idea of DCS [1] is that we can exploit common information during joint reconstruction process to achieve performance improvement, the key of the GDCS framework is that we can exploit not only conventional common information, but also partial common information newly defined in this paper. The proposed GDCS algorithm, therefore, can provide better performance than the DCS algorithm in [1] in a generalized, and practical signal environment.

The remainder of this paper is organized as follows. We summarize the background of CS briefly in Sect. 2. A novel detection algorithm is proposed to capitalize on the GDCS in a practical environment in Sect. 3. In Sect. 4, numerical simulations are provided. Conclusions are made in Sect. 5.

Before going further, some terminologies are clarified as follows.

- Full common information: the set of signal components that are measured by every sensor in a system.
- Partial common information: the set of signal components that are measured by a set of sensor set Π, where its cardinality is $1 < |\Pi| < J$. J is the number of sensors in a system.
- Innovation information: the set of signal components that are measured by a single sensor.
- DCS algorithm: Algorithm presented in [1] to exploit signal structure in the presence of full common information.

2 Compressive Sensing

In many cases, we can represent a real value signal $\mathbf{x} \in R^N$ as sparse coefficients with a particular basis $\Psi = [\psi_1, \ldots, \psi_N]$. We can write

$$\mathbf{x} = \sum_{n=1}^{N} \psi_n \varpi(n) \tag{1}$$

where $\varpi(n)$ is the n th component of sparse coefficients ϖ. Let assume $\|\varpi\|_0 = K$, where $\|\varpi\|_0$ is the number of nonzero elements in vector ϖ. In a matrix multiplication form, it can be represented as

$$x = \Psi \varpi \tag{2}$$

Including the widely used Fourier and wavelet basis, various expansions, e.g., Gabor bases [8] and bases obtained by Principal Component Analysis (PCA) [9], can be used as a sparse basis. For convenience, we use the identity matrix I for a sparse basis $\Psi \mathbb{ZR}$. Without loss of generality, an arbitrary sparse basis can be easily incorporated.

Candes, Romberg, and Tao [10] showed that a reduced set of linear projections can contain enough information to recover a sparse signal, naming this framework as Compressive Sensing (CS). In CS, a compression is simply projecting a signal onto measurement matrix $\Phi \in R^{M \times N}$ where $M < < N$ as follows.

$$\mathbf{y} = \Phi \mathbf{x} \; where \; \mathbf{y} \in R^M \tag{3}$$

This system is ill-posed, however, it can be reconstructed if the restricted isometry property (RIP) of Φ [10] is satisfied with an appropriate constant. According to [10], the original signal $\mathbf{x}\mathbb{RZR}$ can be reconstructed by

$$\varpi_e = \arg\min\|\varpi\|_0 \; s.t. \; \mathbf{y} = \Phi\Psi\varpi \tag{4}$$

However, because of NP-hardness of l_0 minimization, we use l_1 minimization, paying more measurements [10] as a cost of a tractable algorithm.

$$\varpi_e = \arg\min\|\varpi\|_1 \; s.t. \; \mathbf{y} = \Phi\Psi\varpi \tag{5}$$

This approach is called Basis Pursuit. Contrary to l_0 minimization, we can solve l_1 minimization with bearable complexities, which is polynomial in N. In addition to Basis Pursuit, an iterative greedy algorithm can be used for finding the original signal. Orthogonal Matching Pursuit (OMP) [11] is the most typical algorithm.

3   Iterative Signal Detection with Sequential Correlation Search

In this section, we discuss a method that can exploit signal structure without any a priori-knowledge to improve the performance of signal recovery. This is a main obstacle of exploiting partial common information in practical implementation. To compare the requirement of a priori-knowledge of the DCS and the proposed GDCS, the problem formulation of the DCS model [1] is described adopting the same notations.

$$\mathbf{X} = [\mathbf{x}_1^T \, \mathbf{x}_2^T \ldots \mathbf{x}_J^T]^T \in R^{NJ} \tag{6}$$

$$\mathbf{Z} := [\mathbf{z}_C^T \mathbf{z}_1^T \ldots \mathbf{z}_J^T]^T \in R^{N(J+1)} \tag{7}$$

$$\mathbf{x}_j = \mathbf{z}_C + \mathbf{z}_j \ \ where \ j \in \Lambda \tag{8}$$

$$\bar{\Phi} := \begin{bmatrix} \Phi_1 & \Phi_1 & 0 & . & 0 \\ \Phi_2 & 0 & \Phi_2 & . & 0 \\ & . & & . & \\ \Phi_J & 0 & 0 & . & \Phi_J \end{bmatrix} \in R^{JM \times (J+1)N} \tag{9}$$

$$\mathbf{Y} = \bar{\Phi}\mathbf{Z} \tag{10}$$

$$\mathbf{Z}_e = \arg\min \|\mathbf{W}_C \mathbf{z}'_C\|_1 + \|\mathbf{W}_1 \mathbf{z}'_1\|_1 + \ldots + \|\mathbf{W}_J \mathbf{z}'_J\|_1 \ \ s.t. \ \mathbf{Y} = \bar{\Phi}\mathbf{Z}' \tag{11}$$

where $\mathbf{W}_C$ and $\mathbf{W}_j, j \in \Lambda$ are weight matrices, which could be obtained by [12]. Thanks to a joint recovery, improved recovery performance can be obtained compared to disjoint recovery.

Similarly, we can consider a case of the proposed GDCS model, in which a single partial common information is measured by a set of sensors $\Lambda \backslash \{1, 2, 3\}$. This case can be formulated as the following problem by using the proposed GDCS model.

$$\mathbf{X} = [\mathbf{x}_1^T \, \mathbf{x}_2^T \ldots \mathbf{x}_J^T]^T \in R^{NJ} \tag{12}$$

$$\mathbf{Z} := [\mathbf{z}_C^T \mathbf{z}_1^T \ldots \mathbf{z}_J^T]^T \in R^{N(J+1)}, \ where \ \Pi = \Lambda \backslash \{1, 2, 3\} \tag{13}$$

$$\mathbf{x}_j = \begin{cases} \mathbf{z}_{i_j}, & if \, j \notin \Pi \\ \mathbf{z}_{C_\Pi} + \mathbf{z}_{i_j}, & else \end{cases} \tag{14}$$

$$\bar{\Phi} = \begin{bmatrix} 0 & \Phi_1 & 0 & 0 & 0 & . & 0 \\ 0 & 0 & \Phi_2 & 0 & 0 & . & 0 \\ 0 & 0 & 0 & \Phi_3 & 0 & . & 0 \\ \Phi_4 & 0 & 0 & 0 & \Phi_4 & . & 0 \\ . & . & . & . & . & . & 0 \\ \Phi_J & 0 & 0 & 0 & 0 & 0 & \Phi_J \end{bmatrix} \in R^{JM \times (J+1)N} \tag{15}$$

$$\mathbf{Y} = \bar{\Phi}\mathbf{Z} \tag{16}$$

$$\mathbf{Z}_e = \arg\min \|\mathbf{W}_{C_\Pi} \mathbf{z}'_{C_\Pi}\|_1 + \|\mathbf{W}_{i_1} \mathbf{z}'_{i_1}\|_1 + \ldots + \|\mathbf{W}_{i_J} \mathbf{z}'_{i_J}\|_1 \ \ s.t. \ \mathbf{Y} = \bar{\Phi}\mathbf{Z}' \tag{17}$$

where $\mathbf{W}_{C_\Pi}$ and $\mathbf{W}_{i_j}, j \in \Lambda$ are weight matrices, which could be obtained by [12]. As shown above, to exploit partial common information, we have to find the sensor set for partial common information Π, in this case $\Lambda \backslash \{1, 2, 3\}$. Unfortunately, it is not straightforward to find this set. Since each sensor compresses an acquired signal without cooperation of other sensors, there is nothing we can do to determine the correlation structure in a compression process. In a recovery process, although we can find the correlation structure by an exhaustive search, it demands approximately 2^J number of searches, which is not practical. We, therefore, need a moderately complex algorithm that finds the correlation structure.

In this notion, a novel algorithm is proposed for finding a correlation structure, which means a sensor set measuring partial common information. The algorithm iteratively selects the least correlated signal so that we can approximate the sensor set for partial common information Π. For simplicity, we assume a joint sparse signal ensemble $\mathbf{X}$ with partial common information $\mathbf{z}_{C_\Pi}$, where $\Pi = \Lambda \backslash \{1, 2, 3\}$ as in (13). However, since we have no knowledge on the correlation structure, we cannot formulate the measurement matrix as in (15). Instead of that, we refine the correlation structure. Let's assume that the given signal ensemble $\mathbf{X}$ has partial common information, and the correlation structure is not known. We first consider that the given signal ensemble $\mathbf{X}$ has full common information only. Then, the recovery algorithm forcefully makes full common information, while this artificially made full common information is compensated in the innovation information part. For this reason, sensors that do not have partial common information would have more innovation information than the sensors that have partial common information. By using this intuition, we compare l_1 norm of the recovered innovation information part, and remove the sensor whose l_1 norm is maximum from the correlation structure. Repeating this, we can obtain the exact correlation structure.

Although this phenomenon is difficult to understand at the first glance, it is quite straightforward. We should note that the forcefully found full common information may have some relation with the real partial common information. Actually, the forcefully found full common information is likely to be similar to the partial common information to minimize l_1 norm of the solution vector. Then, if the sensor j is one of the sensors that measure the partial common information, a joint recovery process successfully divides the energy of the signal into a joint recovery part (the first column of $\bar{\Phi}$ in (9)) and a disjoint recovery part (the rest of the columns of $\bar{\Phi}$ in (9)). However, if the sensor j, is one of the sensors that do not have the partial common information, the innovation information of the existing DCS for the sensor $j \notin \Pi$ must be made to compensate the forcefully found full common information, causing increase in l_1 norm of the innovation information part.

Thus, it can be exploited only if forcefully found full common information is made to be similar to partial common information. If only a small number of sensors can measure partial common information, i.e., $|\Pi|$ is small, the forcefully found full common information is likely to be different from the partial common information. In this case, we cannot expect to find the sensor set Π based on the above observation. Therefore, in this paper, we assume that any partial common information can be measured by a sufficient number of sensors. This assumption can be justified by the fact that significant performance gain of the proposed GDCS framework can be achieved when a sufficient number of sensors measure partial common information.

Exploiting the above intuition, an iterative signal detection with a sequential correlation search algorithm which we call "GDCS algorithm" throughout this paper is proposed. It is assumed for simplicity that the number of measurements at each sensor is M. The concatenated received signal is denoted by $\mathbf{Y} \in R^{MJ}$. The GDCS algorithm consists of two phases, inner and outer phases respectively. In the inner phase, correlation structure is identified for a given common information by excluding a sensor index one by one from the candidate sensor set. In the outer phase, it determines whether it is going to stop searching a new common information or continue. The details of the proposed algorithm will be elaborated in the paper in preparation for journal publication.

4 Simulation Results

In this section, we demonstrate the GDCS through numerical experiments. Assuming various inter-signal correlations, we compare the detection performance of GDCS algorithm with Oracle-GDCS, which means GDCS with a priori-knowledge of correlation structure the DCS algorithm, and disjoint recovery.

The simulation environment is as follows. Each signal element is generated by an i. i.d. standard Gaussian distribution, and the supports are chosen randomly. The signal size N and the number of sensors J are fixed to 50 and 9, respectively. As aforementioned, the identity matrix is used as a sparse basis without loss of generality.

The measurement matrix is composed of i.i.d. Gaussian entries with a variance $1/M$. We assume a noiseless condition in all simulations. The types of common information and the sparsity of the information are determined as simulation parameters, and the corresponding sensors involved in the correlation are chosen randomly.

We use MATLAB as a simulation tool, and YALL1 solver is used for solving the weighted l_1 minimization. We use an iterative weighted l_1 minimization method introduced in [12] to obtain adequate weight matrices within a reasonable time. The probability of estimation error within the resolution is used as a performance measure where error is calculated by $\|\mathbf{X} - \mathbf{X}_e\|_2 / \|\mathbf{X}\|_2$ and the resolution is set to 0.1.

In Fig. 1(a) and (b), GDCS algorithm outperforms the DCS algorithm when there exists single partial common information while performing as well as oracle-GDCS. It reduces the required number of measurements by 23% and 18%. It is also noted that different performance is due to difference in sparsity of partial common information. We compare the consumed CPU time for GDSS with SCS and the DCS algorithm when the individual number of measurements are 25, 30, and 35. We average the CPU time over 100 different realizations with the simulation setting associated with Fig. 1(a) and (b). The CPU time is measured in seconds. In the (a) environment, the CPU time of the DCS algorithm are 1.42, 1.33, 1.32, respectively, while those of GDCS with SCS are 4.07, 3.92, 3.89, respectively. In the (b) environment, the CPU time of the DCS algorithm are 1.10, 1.08, 1.12, respectively, while those of the GDCS algorithm are 3.52, 3.30, 3.36, respectively. We can observe that the GDCS improves the performance with marginal increase in CPU time.

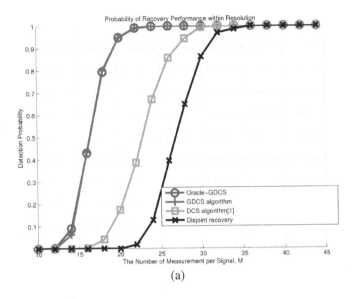

(a)

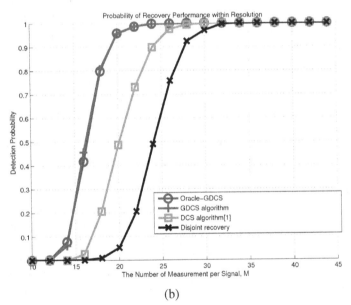

(b)

Fig. 1. Performance comparison of Oracle GDCS, GDCS algorithm, DCS algorithm and disjoint recovery when (a) $K_{i_j} = 4$, $K_{C_\Pi} = 6$, $|\Pi| = 6$, (b) $K_{i_j} = 4$, $K_{C_\Pi} = 4$, $|\Pi| = 6$

5   Conclusions

In this paper, we proposed a new framework, a generalized version of the conventional one [1] so that it can be applicable to a more realistic environment. The proposed GDCS model refines the existing model so that it can exploit signal structure associated with partial common information in the joint recovery process. In this notion, we proposed GDCS algorithm to exploit this information in joint signal recovery without a priori-knowledge. Numerical simulation verifies that the proposed algorithm can reduce the required number of measurements compared to the DCS algorithm.

Acknowledgement. This research was supported by Basic Science Research Program through the National Research Foundation of Korea(NRF) funded by the Ministry of Science, ICT & Future Planning (2015R1C1A1A02037515), and (2012R1A2A2A01047554).

References

1. Baron, D., Duarte, M.F., Wakin, M.B., Sarvotham, S., Baraniuk, R.G.: Distributed compressive sensing, arXiv.org, vol. cs.IT, January 2009
2. Tropp, J.A., Gilbert, A., Strauss, M.: Simultaneous sparse approximation via greedy pursuit. In: Proceedings of the IEEE International Conference on Acoustics, Speech, and Signal Processing, pp. 721–724, March 2005
3. Davies, M., Eldar, Y.: Rank awareness in joint sparse recovery. IEEE Trans. Inform. Theory **58**(2), 1135–1146 (2012)
4. Tao, G.Z.Y., Zhang, J.: Guaranteed stability of sparse recovery in distributed compressive sensing MIMO radar. Int. J. Antenna Propag. **2015**, 10 (2015)
5. Chen, M.R.D.R.W., Wa, I.J.: Distributed Compressive Sensing Reconstruction Via Common Support Discovery. In: Proceedings of the IEEE International Conference on Communications, pp. 1–5 (2011)
6. Caione, D.B.C., Benining, L.: Compressive sensing optimization for signal ensembles in WSNs. IEEE Trans. Industrial Info. **10**(1), 382–392 (2013)
7. Singh, A., Dandapat, S.: Distributed compressive sensing for multichannel ECG signals over learned dictionaries. In: Proceedings of INDICON, Pune, pp. 1–6 (2014)
8. Mallat, S.: A wavelet tour of Signal Processing: The Sparse Way, 3rd edn. Academic Press, London (2008)
9. Masiero, R., Quer, G., Munaretto, D., Rossi, M., Widmer, J., Zorzi, M.: Data acquisition through joint compressive sensing and principal component analysis. In: Proceedings of the IEEE Globe Telecom Conference, pp. 1–6, November 2009
10. Candes, E., Romberg, J., Tao, T.: Robust uncertainty principles: exact signal reconstruction from highly incomplete frequency information. IEEE Trans. Inform. Theory **52**(2), 489–509 (2006)
11. Tropp, J.A., Gilbert, A.: Signal recovery from random measurements via orthogonal matching pursuit. IEEE Trans. Inform. Theory **53**(12), 4655–4666 (2007)
12. Dai, W., Milenkovic, O.: Subspace pursuit for compressive sensing signal reconstruction. IEEE Trans. Inform. Theory **55**(5), 2230–2249 (2009)

MMSE Based Interference Cancellation and Beamforming Scheme for Signal Transmission in IoT

Xin Su[1], YuPeng Wang[2(✉)], Chang Choi[3], and Dongmin Choi[4]

[1] Changzhou Key Laboratory of Robotics and Intelligent Technology,
College of IoT Engineering, Hohai University, Changzhou, China
leosu8622@163.com
[2] College of Electronic & Information Engineering,
Shenyang Aerospace University, Shenyang, China
ypwang@sau.edu.cn
[3] Department of Computer Engineering, Chosun University, Gwangju, Korea
enduranceaura@gmail.com
[4] Division of Undeclared Majors, Chosun University, Gwangju 61452, Korea
jdmcc@chosun.ac.kr

Abstract. Due to the application nature of Internet of Things, various different kinds of devices will be connected to the network. To increase the access opportunity, a spreading based signal transmission technique is proposed to be used in the signal transmission of IoT. To alleviate the effects of rich scattering transmission environment and interferences from other devices, MMSE based interference cancellation combined with the receive beamforming is proposed in this paper. Through simulation, we observe that the proposed algorithm is outperforms the conventional spreading transmission schemes and is suitable for the transmission environment of IoT.

Keywords: IoT · Interference cancellation · Receive beamforming · MMSE

1 Introduction

As the rapid development of science and technology, the world is becoming smarter and smarter. In the smart word, people have extended their connection needs from the hand phone, PDA, notebook to the smart wearable devices, smart transportations, smart environment, and etc. Eventually, all the network world, physical world, and social world which are related to our daily life will be merged together to form a highly inter-connected smart world. The IoT (Internet of Things) technique which is considered as the corner stone of the future smart world, is continuously taking huge attentions from academia and industry [1–3].

Currently, the research around IoT focuses on the high layer issues, networking issues, and energy harvesting issues. Maria Gorlatova proposed a method which evaluates the energy recovery efficiency based on the trajectory of acceleration speed and applied the related algorithm in the design of energy recuperator [4]. Xiaosen Liu proposed an energy harvesting method based on 0.18 mm COMS [5]. Zhiqui Chen

© ICST Institute for Computer Sciences, Social Informatics and Telecommunications Engineering 2017
J.J. Jung and P. Kim (Eds.): BDTA 2016, LNICST 194, pp. 138–143, 2017.
DOI: 10.1007/978-3-319-58967-1_16

proposed to utilize the device's remaining energy in the routing method to optimize the transmission energy efficiency [6]. However, little research was done on the physical layer signal transmission. In this paper, we introduce the concept of spreading and smart antenna to mitigate the effects of rich scattering transmission environment and interference among IoT users.

The remaining part of this paper is as follows. In Sect. 2, we formulate the physical layer transmission problem considered in this paper. The proposed IoT signal transmission technique is described in Sect. 3. The evaluation results are given in Sect. 4. Conclusions are drawn in Sect. 5.

2 Problem Formulation

The IoT is composed of many different kinds of devices to satisfy people's various daily needs. In order to alleviate the communication coordination environment and provide more medium access opportunities, the concept of spectrum spreading is introduced in this paper. By selecting the proper pseudo-random sequence such as m-sequence or gold sequence, the interference between difference devices can be mitigated.

The transmitted signal of the kth user after the spreading operation $x_k(t)$ is shown in (1).

$$x_k(t) = \sqrt{E_k} g_k(t) b_k(t) \tag{1}$$

where $\sqrt{E_k}$ is the amplitude of the transmitted signal; $g_k(t)$ is the pseudo-random sequence to spread the transmitted signal; $b_k(t)$ is the transmitted data symbol;

After de-spreading operation, the kth user's received signal is as

$$
\begin{aligned}
y_k(t) = {} & h_k \frac{1}{N} \sum g_k(t) A_k(t) g_k(t) b_k(t) \\
& + \sum_{j=1, j \neq k}^{K} h_{j,k} g_k(t) A_k(t) g_k(t) b_k(t) \\
& + n_k(t)
\end{aligned}
\tag{2}
$$

where h_k represents the channel attenuation; $h_{j,k}$ represents the interfering channel attenuation. The first part of the (2) represents the desired received signal and the second part represents the interference from other devices.

3 Proposed MMSE Based IoT Transmission Technique

In this paper, we utilized the minimum mean square error (MMSE) criterion to mitigate the effects of channel attenuation and interference. Firstly, we adopt the MMSE multi-user detection method to eliminate the interference from other users. Then, MMSE based receive beamforming technique is used to obtained the directivity gain of the antenna array.

3.1 MMSE Based Interference Cancellation

To minimize the mean square error, the MMSE based interference cancellation method is to find a optimum weight $\mathbf{L}_{mmse}$, which satisfies (3) [7].

$$\mathbf{L}_{MMSE} = \arg\min_{L}\left\{ E\|\mathbf{b} - \mathbf{L}\mathbf{y}\|^2 \right\} \tag{3}$$

To make $\frac{d}{dL}\left[E\|\mathbf{b} - \mathbf{L}\mathbf{y}\|^2 \right] = 0$, we obtain

$$\mathbf{L}_{MMSE} = [\mathbf{R} + \sigma^2 \mathbf{A}^{-1}]^{-1} \tag{4}$$

where $\mathbf{R}$ is the cross-correlation matrix of the pseudo-random sequence; $\mathbf{A} = diag(\sqrt{E_1}, \sqrt{E_2}, \ldots, \sqrt{E_K})_{KxK}$; σ^2 is the variance of AWGN noise.

3.2 MMSE Based Receive Beamforming

To minimize the effects of channel attenuation, smart antenna technique is adopted at the receiver side to obtain the temporal and spatial focusing property, which is provided by the directivity gain of the antenna array. In the proposed algorithm, MMSE based weight calculation method is used. The cost function to be minimized is similar as (3) [8].

$$\mathbf{W}_{MMSE,k} = \arg\min_{W}\left\{ E\|\mathbf{b} - \mathbf{W}\mathbf{y}'_k\|^2 \right\} \tag{5}$$

where $\mathbf{y}'_k$ is the output matrix after the MMSE based interference cancellation of the kth user.

Solving (5) gives the expression for the optimum weights based on MMSE for the kth user as (6)

$$\mathbf{W}_{MMSE,k} = \mathbf{R}_{y'_k y'_k}^{-1} \mathbf{r}_{y'_k d} \tag{6}$$

where $\mathbf{R}_{y'_k y'_k}$ is the covariance matrix of the kth user's received signal after the interference cancellation; $\mathbf{r}_{y'_k d}$ is the cross correlation matrix between the signal y'_k and pilot signal d.

Figure 1 shows the detailed structure of the proposed signal transmission technique.

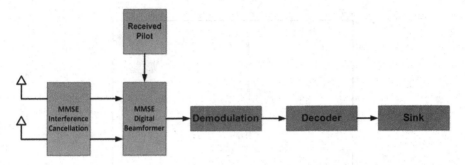

Fig. 1. Detailed signal processing procedure of the proposed algorithm.

4 Performance Evaluation

4.1 Simulation Parameters

In this section, we evaluate the proposed algorithm by the Monte-Carlo simulation. To consider the rich scattering transmission environment of IoT, SCM channel model is utilized instead of the normal Rayleigh fading channel. The detailed simulation parameters are shown in Table 1.

Table 1. Detailed simulation parameters.

Parameter	Value
No. of Devices	8
FEC	Convolutional Code
Modulation	QPSK
PN Sequence	M-sequence
Channel Model	SCM
SIR for Interference	0 dB
No. of Antennae	2 for Rx

4.2 Simulation Results

Figure 2 shows the temporal focusing property of the proposed signal transmission technique. To represent the transmission delay, we insert zeros at the beginning of the transmission data sequence. From the figure, we observe only one single peak at the timing where the real data is received, which shows a good timing focusing property.

The BER performance comparison among single antenna spreading transmission without interference cancellation, single antenna spreading transmission with interference cancellation and the proposed algorithm is shown in Fig. 3. From the figure, we found that the spreading operation has little effects in the rich scattering environment, where the received signal suffers heavy channel attenuation and interference. Even though the MMSE based interference cancellation method is used, the signal still have a very high error ratio. However, the proposed algorithm works better than the other

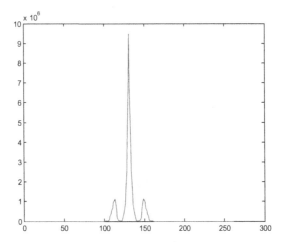

Fig. 2. Temporal focusing property of the proposed algorithm.

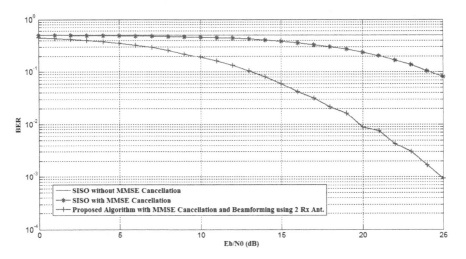

Fig. 3. BER comparison among the proposed algorithm and conventional algorithms

two conventional algorithms due to the cancellation on the interference and elimination on the channel attenuation, the channel attenuation and interference are eliminated.

5   Conclusions

In this paper, we proposed a MMSE based signal transmission scheme to mitigate the effects of rich scattering environment and interference from other devices. From the evaluation results, we found that the proposed algorithm is effective on the interference mitigation and bit error rate reduction.

Although the MMSE based criterion has some application limitations such as the prior knowledge on the channel quality, this can be easily achieved by specific pilot design, which is out of scope of this paper. In addition, most devices served by IoT are an actually low-speed terminal, which means high density pilot insertion is not required to track the channel variations.

Acknowledgements. This research is supported by the research grant (with No. of L2014067) of Liaoning Provincial Education Department of China and in part by the Natural Science Foundation of Jiangsu Province under Grant BK20160287, and also in part by Fundamental Research Funds for the Central Universities under Grant 2015B30614.

References

1. Ashton, K.: That 'Internet of Things' thing in the real word, things matter more than ideas. RFID J. **22**, 97–114 (2009)
2. Brock, D.L.: The electronic product code (epc) a naming scheme for physical objects. Auto-ID Center, Whiter Paper, January 2001
3. Kortuem, G., Kawsar, F., Fitton, D., Sundramoorthy, V.: Smart objects as building blocks for the internet of things. IEEE Internet Comput. **14**(1), 44–51 (2010)
4. Gorlatova, M., Sarik, J., Grebla, G., et al.: Movers and shakers: kinetic energy harvesting for the internet of things. IEEE JSAC **33**, 1624–1639 (2015)
5. Liu, X., Sánchez-Sinencio, E.: A highly efficient ultralow photovoltaic power harvesting system with MPPT for internet of things smart nodes. IEEE Trans. VLSI Syst. **23**, 3065–3075 (2015)
6. Chen, Z., Ni, J., Jiang, G., Liu, C.: Load balanced routing protocol based on the time-domain energy level in WSN. Microelectron. Comput. **27**(12), 83–86 (2010)
7. Zhang, X.D., Wei, W.: Blind adaptive multiuser detection based on Kalman filtering. IEEE Trans. Signal Process. **50**(1), 87–95 (2002)
8. Jafar, S.A., Vishwanath, S., Goldsmith, A.: Channel capacity and beamforming for multiple transmit and receive antennas with covariance feedback (2001)

A Multi-user-collaboration Platform Concept for Managing Simulation-Based Optimization of Virtual Tooling as Big Data Exchange Service

An Implementation as Proof of Concept Based on Different Human-Machine-Interfaces

Jens Weber[(✉)]

Heinz Nixdorf Institute, Fuerstenallee 11, Paderborn, Germany
Jens.weber@hni.upb.de

Abstract. Intelligent connected systems for the successfully implementation of collaborative work systems in the areas such as Internet of things/Industry 4.0 require a knowledge management system which offers opportunities to work on one task with different organizations on the same time. Cooperative work in the field of setup preparation for production systems is one challenge for an efficiency and infallibly work preparation. One example is the validation and optimization process for NC-programs, which is offered by CAD/CAM interfaces as well as the experiences of the worker uses the machine. Planned production processes are simulated by the CAD/CAM-programs. Optimized setup data are provided to the worker using the setup. The challenge is to provide a service to handle the dataset of job information, optimization information and setup information for many users in order to manage databases and data sets. The approach deals with a system concept of an implementation of a production optimization tool embedded by a collaborative platform containing access by a Multi-User-Agent to manage setup parameters direct from the simulation to the machine as well as proved job management workflows.

Keywords: Simulation-based optimization · Collaborative platform · Multi-user-agent · Knowledge management · Virtual tooling · NC-program · Data exchange

1 Introduction

Intelligent linked systems are an important condition in order to implement research activities and manufacturing jobs in the area of "internet of things" – in Germany often called "Industry 4.0". Thus a qualified knowledge management system is required which offers opportunities to work synchronously and cooperatively on one task on the same time independent from the workplace location. The goal is to reach a productive and sustainable manufacturing process which offers also the opportunity to learn from experience and data from the past.

Thus a Multi-User-Agent based approach is pursued which provides work preparation processes in the area of virtual tooling as part of the digital factory. In addition,

© ICST Institute for Computer Sciences, Social Informatics and Telecommunications Engineering 2017
J.J. Jung and P. Kim (Eds.): BDTA 2016, LNICST 194, pp. 144–153, 2017.
DOI: 10.1007/978-3-319-58967-1_17

cooperative work is also provided using a collaborative platform. With support of a combination of several systems and approaches, a service platforms is developed and will be presented in this contribution. For that knowledge, production and setup data sets are provided for several users. Especially for the generation of job and order data, production details, setup information for workflows and tooling machine setups as well as the machine-generated sensor data offers new challenges for process analysis and data processing. A detailed problem description is given in Sect. 2. Section 3 presents the related work for the basic research project as well as collaborative systems. Section 4 offers the current concept and the architecture of the total work preparation system. Section 5 closes the contribution with a conclusion.

2 Problem Description and Motivation

In order to provide for each user $u_m \in \{u_1 \ldots u_M\}$ inside and outside of several organizations $o_n \in \{o_1 \ldots o_N\}$, every manufacturing job are defined as $j_k \in \{j_1 \ldots j_K\}$. As cloud and collaborative application, these set would rapidly lead to high data variants that have to be evaluated by optimization and simulation runs as well as job scheduling queues. Each job contains at least one workpiece. The job information are provided as instructed data which are saved in a database. These data are processed in a structured data format, the so called VMDE. VMDE stands for virtual machine data exchange and is defined as an XML-based data format (see Fig. 1) which provides the definition of a job for the tooling machine simulation. The XML-based data contains tool information, workpiece geometry information, workpiece clamp information, position information, encoded NC-programs as well as control information. The NC-program, control information and position information defines the setup information that can be optimized by the "setup optimizer".

```
<VMDE>                                    <WorkpieceClamp id="WC1000" />
    <Head />                                  <Workpiece id="1" name ="square">
    <Project>                                 <zeropoint id="1" location="-70 30
    <Tool id="T1000">                         25"/>
    <Clamp>                                   ........
    <!—Geometry content -->                   </Workpiece>
    <Blade>                                    <Programs id="26"/>
    <!—Geometry content -->                   </Project>
    <Holder>                               </VMDE>
    <!—Geometry content -->
    </Holder>
    </Blade>
    </Clamp>
    </Tool>
........
```

Fig. 1. Example interface data for communication between simulation and work preparation platform

Clamp and workpiece geometry (CAD-data) depends on the design process of the organization. The problem for these circumstance is that one production step can be defined as one manufacturing job it and can contain one and/or more VMDE-data. These configurations lead to a data processing overhead. In order to provide a processing method to define jobs, VMDE contents in a cloud-based collaborative environment, a process system is striven by this contribution. For that a standardized workflow is required to organize collaboration between the organizations and their jobs.

A rough example structure of the communication interface (VMDE) between work preparation platform and simulation is shown in Fig. 1.

In order to provide the manufacturing job configuration (implemented as VMDE-data) which contains always a simulation job as available for lower system performance, a subdivision of the data in smaller string-snippets are required. Then the data process is provided as well running system using the cloud architecture. The snippets are saved in the provided individual user database. Figure 2 shows a database condition for a simple simulation job in order to simulate primitive geometry. For more complex jobs, the data number will be rapidly increase which requires high data processing procedures. For one simple job which contains only one simulation job-data in order to communicate with the simulation model, the number of snippets amount 20094 strings.

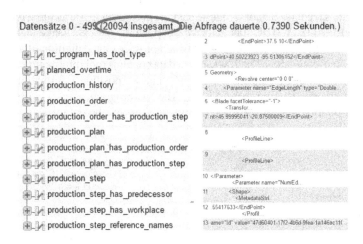

Fig. 2. Section from one user database containing simulation job information and string-snippets

3 Related Work

3.1 Related Work of the Development Process of the Work Preparation Platform

The basic research project in order to improve work preparation-processes in the area of manufacturing processes especially cutting machines named by "*InVorMa*" is supported by the German Ministry of Education and Research. The research goal is to develop a service platform in order to optimize machine and job setups based on virtual tooling using a simulation model of a milling machine. The research project contains sub-projects which handle the setup optimization processes as well as the knowledge management system. The result is an optimal job schedule of manufacturing and simulation jobs, providing valid setup parameters, suitable machine selection and a suitable data processing.

For that there arose several approaches to decrease the simulation and computational effort. The computational process occurs because of the high number of variants of the machine setup and production parameter setup. The optimization would generate many parameter sets for each product and each machine which has to be simulated. The idea to parallelize the simulation would imply a linear time decreasing situation which offers useful improvements for low number of datasets.

The optimization component of the production and machine setup is given by a metaheuristic called particle-swarm optimization (PSO) (see [1]) which is tested as asynchronous, synchronous, and grouped-asynchronous extension in order to manage stochastic node failures and parallelize optimization runs [2]. A shrank process time using asynchronous PSO is shown in this contribution. In combination with an additional calculation program called "*NC-parser*" as fitness component the processing time decreases on several seconds. The NC-parser approximates the tool paths and estimates the production time using a real NC-program. In combination with the PSO-algorithm minimal workpiece positions are identifiable which leads minimal production time. In the contribution of Weber [3] zeropoint optimization of the workpiece setup in order to reach minimal production time is presented.

In order to concern the problem of high simulation effort cluster algorithms are tested in combination with the PSO-NC-Parser-combination. The contributions [4–6] offer a useful solution in order to optimize workpiece and clamp position for the machine setup to determine positions without unintentional collisions during the production processes. The collected positions information are also saved back in the user data bases which are managed by a multi-user-agent.

3.2   Related Work to Collaborative Platforms and Conceptual Overview

For systems which offer integrated and collaborative work for several participants, for example customer and business partners, collaborative systems are defined as "computer based systems for group oriented support of decisions support and problem solutions, flexible information link and exchange, rapid dialogues including absence of participants, electronic conferences and worldwide multi-media communication [7]." This kind of systems represents no new approaches, but due to the research in order to implement support opportunities for cooperative teams using information and communication technologies, the research idea arose in the early 1980s as "*Computer Supported Cooperative Work (CSCW)*." It is represented as interdisciplinary research area in order to improve team work providing and usage suitable information and communication systems [7]. Collaborative systems gain increasing importance caused by the provided internet connection, networks in companies, and the development of new devices though to the current system landscape today [8]. With support of web services and defined interfaces, organizations can disclose, receive, and share information with their cooperative units. CSCW is designated as predecessor of Social Software/Social Media [9, 10]. A conceptual map is presented in the contribution of Martensen [9].

4   Concept and System Architecture

4.1   System Architecture

The work preparation platform considers five subsystems in order to provide optimal tooling jobs and machine setup. The usability is realized by a multi-user-login structure which is controlled by a web-based interface and also by a collaborative platform. Figure 3 presents the sub-systems embedded in a cloud-architecture including the interfaces between users, system, and manufacturing part.

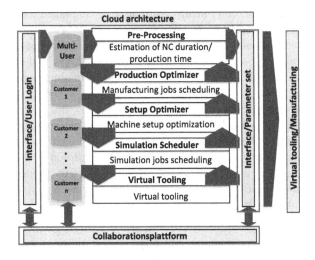

Fig. 3. System architecture of the work preparation platform

The user gets their individual database connection by a user log-in so that their have access to their own system resource. With support of the web-based interface as well as the collaborative platform, there are opportunities to define the manufacturing jobs with meta-data such as deadlines, costs, setup costs, selections of workpiece geometries, workpiece clamps as workpiece fixtures in the tooling machine, the used machine, chosen NC-programs and trigger of simulation and optimization of their schedule and machine setup.

The pre-processing part uses the *NC-parser* application, a program which read in the NC-commands and approximates the tool paths and calculate the production time, and in combination with an ontology system a suitable machine can be chosen taking into account the workpiece size, tool paths as boundary box, and the workspace size of the potential tooling machine.

The production optimizer calculates a manufacturing job schedule taking into account the restrictions such as machine capacities, human resources, job and machine failures, time tables and delivery deadlines.

The setup optimizer uses a meta-heuristic as optimization part which is combined with a cluster algorithm (see [2, 3, 5]) in order to decrease the calculation effort and the number of simulation runs when all potential solutions have to be evaluated by the simulation programs. The goal of the setup optimizer is to find a best position and orientation of workpiece and clamps in the machine workspace. The potential solution candidates from the setup optimizer system are distributed as simulation jobs to computer resources by the simulation scheduler. Then the solution candidates are evaluated by the virtual tooling system which contains the simulation model of a real tooling machine. There, the workpiece position candidates are investigated in order to prevent unintentional collisions as well as circumstantial tool paths.

The results of the optimization systems are given back to the user by the web-interface as well as the collaborative platform and the results are also stored in the database device where the user has access for future processes.

4.2 Multi-user-agent and Interface Concept

As described in Sect. 3.1 the work preparation platform consists of several parts which are joined together and interact with many users in- and outside of organizations. The focus of this contribution is the user interface by collaborative platform and web-interface provided by database resources and multi-user-agent. A detailed content of the subsystems of the work preparation platform is given by the references in Sect. 2.

Figure 4 gives an overview about the interaction of the subsystems as well as the part of the user log-in and the interfaces to the collaborative platform.

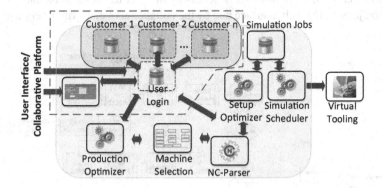

Fig. 4. System interactions and embedded multi-user-agent and collaborative platform (Color figure online)

The red marked field in Fig. 4 presents the system overview which communicate with the user database where the user access is provided. The multi-user-agent communicates with the sub-systems of the work preparation platform depending on the intention of the user (optimization, job-scheduling, simulation). The data sets of the user which are inserted in the interfaces lead to a selected machine, a production plan and an individual machine setup. Every user uses different tooling machines which are provided as simulation model. The optimization can be calculated in their assigned network using assigned computer resources or are provided by a cloud architecture.

The virtual tooling data exchange process is controlled by the simulation scheduler as well as the remaining systems using the cloud architecture which is presented in Sect. 4.1. The size of the exchange data volume depends on the number of users, complexity of the workpieces and the total number of simulation jobs and can determine a range size from several Megabytes to several Gigabytes for each job.

The multi-user-agent consists of a so called customer-pool database where all user names, passwords and related server access are saved. In order to ensure the personal user-data, the entries are encoded as "base64" containing ASCII characters. The entries are additionally secured by secure-hash-algorithm ("sha 512") which process cryptology hash functions. In order to manage the multi-user-agent and the databases as well as the total number of jobs, there is a workflow management system required which is realized by a HMI (human machine interface) provided as web interface as well as collaborative platform. As use case the commercial platform Microsoft SharePoint is used. The web interface is programmed by the standard-tools HTML 5 and the database connection is managed by PHP 5. The features of the interface is realized by JavaScript. The data bases are implemented as SQL-database, but alternative database implementation opportunities are conceivable.

4.3 Workflow as Online Solution to Define Simulation Jobs and Production Information in Order to Manage Work Preparation

Figure 5 presents the workflow that is necessary in order to define full simulation and production jobs by the web interface and by the collaborative platform. Every user uses

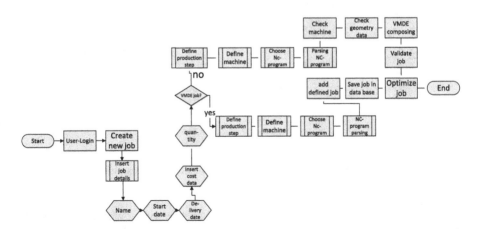

Fig. 5. Workflow of the data management process using web interfaces as well as collaborative platform

his database connection and the services behind the optimization as well as the setup check. The VMDE-composing process is also implemented as a work preparation platform. The Optimization and the composer processes in order to manage the big data problem caused by the VMDE-snippets are independent.

4.4 Online Solution Using Collaborative Platform

The web interface, presented in Sect. 4.3 is recreated as collaborative platform for MS SharePoint. The user can define simulation jobs, manage production information as well as organize the setup of the work preparation issues. The database is also linked with the provided server architecture. In order to run optimization processes the collaborative platform is less well suitable because the optimization executed data has to be saved on the local system of the user. But for the workflow management and for the data management in order to provide simulation jobs and resource management, the platform offers its useful features from different log-in locations. The interface design is shown in Fig. 6.

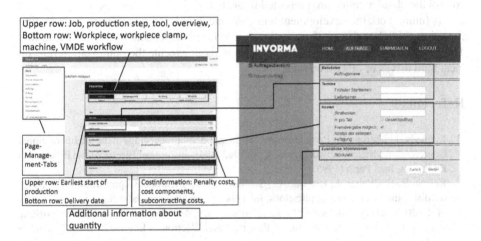

Fig. 6. Collaborative interface using MS SharePoint as use case and proof of concept

The data set is stored in lists and as entries in the database and it can be read-out by standard SQL-commands. Behind the detailed information about the job, there are entities such as name and quantity. Production steps contains the information about technology such as milling, drilling or turning or a combination as well as additional activities such as assembly, painting or deburring.

The input mask "tool" defines the tool geometry and the tool type. Information about workpiece and clamp contain material type, names and definitions and geometry data. The VMDE workflow contains additional information that are required to use a composer application in order to create the VMDE data required for the data exchange with the virtual tooling machine, for example a milling machine simulation model. The collaboration platform also manages the delivery dates, production dates and shift schedules for the worker and the combined costs. A decision support function is also possible through the information about the subcontract information as well as the external processing costs.

5   Conclusion and Outlook

This contribution represents a symbiosis of technical opportunities to combine and to synchronize a web-based multi-user-interface and a collaborative platform for work preparation management of many generated optimization data provided by the tooling machines (sensor data), production optimizer as well as setup optimizer and the virtual tooling machine data exchange process (VMDE). Especially the high number of data exchange snippets (VMDE-snippets) in combination with sensor data cause a bottle neck in order to manage the data set for the o_n organizations that contains u_m users and j_k jobs. The provided platform shows a high potential of usability and guarantees an overview about the simulation jobs and organize the validation and optimization runs in order to control the cloud architecture presented in Sect. 4.

As future work the development will continue in order to synchronize optimization runs in the collaborative platform and web-interface independent of the login-location. The goal is to use the collaborative platform without execute the optimization runs on local hard drives or other hardware.

References

1. Kennedy, J., Eberhart R.: Particle swarm optimization. In: Proceedings of the 4th International Conference on Neural Networks, vol. 4, pp. 1942–1948. Institute of Electrical and Electronics Engineers, Inc., Perth (1995)
2. Reisch, R.-E., Weber, J., Laroque, C., Schröder, C.: Asynchronous optimization techniques for distributed computing applications. In: Tolk, A., Padilla, J. J., Jafar S. (eds.) Proceedings of the 2015 Spring Simulation Multi conference 2015, 48th Annual Simulation Symposium, vol. 47, no. 2, pp. 49–57. Institute of Electrical and Electronics Engineers, Inc., Alexandria (2015)
3. Weber, J.: A technical approach of a simulation-based optimization platform for setup-preparation via virtual tooling by testing the optimization of zero point positions in CNC-applications. In: Yilmaz, L., Chan, W.K.V., Moon, I., Roeder, T.M.K., Macal, C., Rossetti, M.D. (eds.) Proceedings of 2015 Winter Simulation Conference, Huntington Beach (2015)
4. Weber, J., Mueß, A., Dangelmaier, W.: A simulation based optimization approach for setting-up CNC machines. In: Doerner, K.F., Ljubic, I., Pflug, G., Tragler, G. (eds.) Operations Research Proceedings 2015. ORP, pp. 445–451. Springer, Cham (2017). doi: 10.1007/978-3-319-42902-1_60
5. Mueß, A., Weber, J., Reisch, R.-R., Jurke, B.: Implementation and comparison of cluster-based PSO extensions in hybrid settings with efficient approximation. In: Niggemann, O., Beyerer, J. (eds.) Machine Learning for Cyber Physical Systems. TIA 2016, vol. 1, pp. 87–93. Springer, Heidelberg (2016). doi:10.1007/978-3-662-48838-6_11

6. Laroque, C., Weber, J., Reisch, R.-R., Schröder, C.: Ein Verfahren zur simulationsgestützten Optimierung von Einrichtungsparametern an Werkzeugmaschinen in Cloud-Umgebungen. In: Nissen, V., Stelzer, S., Straßburger, S., Firscher, D. (eds.) Proceedings of the Multikonferenz Wirtschaftsinformatik (MKWI 2016), edited by V. Nissen, D. Stelzer, S. Straßburger, D. Firscher, vol 3. pp. 1761–1772. Monsenstein und Vannerdat OHG, Universitätsverlag Ilmenau, Ilmenau (2016)
7. Wendel, T.: Organisation der Teamarbeit. Betriebswirtschaftlicher Verlag Gabler GmbH, Wiesbaden (1996)
8. Riemer, K., Arendt, P., Wulf, A.: Marktstudie Kooperationssysteme. Von E-Mail über Groupware zur Echtzeitkooperation. Cuvillier Verlag, Göttingen (2009)
9. Mertensen, M.: Einsatz von Social Software durch Unternehmensberater, Akzeptanz, Präferenzen, Nutzungsarten. Gabler Verlag, Springer Fachmedien, Wiesbaden (2014)
10. Back, A, Gronau, N., Tochtermann, K. (eds.): Web 2.0 und Social Media in der Unternehmenspraxis. Grundlagen, Anwendungen und Methoden mit zahlreichen Fallstudien. Oldenbourg Wissenschaftsverlag GmbH, München (2012)

Author Index

Bae, Kitae   130

Cho, Kyu Cheol   40
Choi, Chang   138
Choi, Dongmin   138
Cuomo, Salvatore   54

De Michele, Pasquale   54

Furukawa, Yuki   23

Gullo, Francesco   54

Han, Young Shin   40
Hayano, Junichiro   23
Hong, Jaemin   29
Hong, Jiwon   122
Hong, Taekeun   83
Hwang, Dosam   48, 64
Hwang, Seunggye   130

Jung, Jason J.   48, 64

Kim, ChongGun   29
Kim, Dong Ku   130
Kim, Jae Kwon   40
Kim, Jeongin   83
Kim, Pankoo   83
Kim, Sang-Wook   122
Ko, Hoon   130
Koptyra, Katarzyna   115

Lee, Eunji   83
Lee, Jong Sik   40
Lee, Kyungroul   105
Lee, O.-Joun   48

Mukherjee, Saswati   12

Nagarajan, Shivanee   12
Nguyen, Tuong Tri   64

Ogiela, Marek R.   115
Oh, Insu   105

Park, Jeonghun   130
Park, Sanghyun   122
Piccialli, Francesco   54
Ponti, Giovanni   54
Poulose Jacob, K.   3
Prasomphan, Sathit   73

Ramesh, Varun   12

Su, Xin   138

Tagarelli, Andrea   54
Tran, Quang Dieu   48

Unnikrishnan, A.   3

Varghese, Bindiya M.   3

Wang, YuPeng   138
Weber, Jens   91, 144
Wu, Mary   29

Yang, Janghoon   130
Yim, Kangbin   105
Yoshida, Yutaka   23
Yuda, Emi   23

Printed in the United States
By Bookmasters